Research and Perspectives in Longevity

Springer

Berlin
Heidelberg
New York
Barcelona
Budapest
Hong Kong
London
Milan
Paris
Santa Clara
Singapore
Tokyo

J.-M. Robine J.W. Vaupel
B. Jeune M. Allard (Eds.)

Longevity:
To the Limits and Beyond

With 51 Figures and 31 Tables

 Springer

Jean-Marie Robine
Equipe INSERM Démographie et Santé
Val d'Aurelle
Parc Euromédecine
34298 Montpellier cedex 5
France

James W. Vaupel
Odense University
Centre for Health and Social Policy
Winslowparken 17
5000 Odense C
Denmark

Bernard Jeune
Odense University
Centre for Health and Social Policy
Winslowparken 17
5000 Odense
Denmark

Michel Allard
Fondation IPSEN
24, rue Erlanger
75781 Paris cedex 16
France

ISBN 3-540-62945-9 Springer-Verlag Berlin Heidelberg New York

Herstellung: PRO EDIT GmbH, D-69126 Heidelberg
Cover design: Design & Production
Typesetting: Mitterweger Werksatz GmbH, 68723 Plankstadt, Germany
SPIN: 10628981   27/3136/SPS – 5 4 3 2 1 0 – Printed in acid-free paper

Preface

Why longevity?

For a number of years, the Fondation IPSEN has been devoting considerable effort to the various aspects of ageing, not only to age-related diseases such as Alzheimer's, but also to the Centenarians, the paragon of positive ageing. The logical continuation of this approach is to address the question of longevity in global terms. Behind the extreme values, what span is accessible to all of us and likely to directly concern most of our contemporaries? The individual and collective increase in the duration of life is one of the most striking phenomena of our time. It could be one of the most significant events in the "bio-social" history of humanity. The increase in life expectancy at old age, which started a few decades ago only, is going on. The most well-advised observer had not foreseen or even dared hope for this increase which will drastically affect our everyday life, our habits and our behavior. In the fragment of human history we are living in, it is our responsibility to deal with this major transformation for the species. Such a transformation needs an effort from all to adapt to the new conditions. This transformation has to be managed rather than simply experienced, anticipated rather than followed, in order to avoid any attempt to pervert this major step forward.

All that was present during the first symposium of the new series on longevity of the *Colloques Médecine et Recherche* convened by the Fondation IPSEN. Although demography, biology, genetics and many other disciplines will be involved in these series, future symposia will lay particular emphasis on one specific aspect, as usually for these *Colloques*.

Longevity: to the limits ... and beyond

The title sums up the subject well: Up to what point will life expectancy increase? And then: what? Is there a theoretical or practical limit to human longevity and does this shift or not? Is it utopian to envisage a life expectancy of 100 years or more at birth? Is the emergence of centenarians recent or not? What model can be used to study longevity? What are the biological determinants of this current change? Why and how does this change differ between individuals, between sexes, between countries, between areas? What is the *"raison d'être"* of this transformation with regard to the human species evolution? What are the respective roles of genetics and environment? Under which conditions and at what price are these years of life gained? The list of questions is unending. However, we hope to have answered some of them in this book, presenting the papers of the first *Colloques Médecine et Recherche* of Fondation IPSEN on longevity.

The logo of this series, a symbolic tree represented on the shell scales of a stylised tortoise, is appropriate for putting across a metaphor which no-one will fail to understand: slow certainly, but likeable and determined, exactly like our approach and the duration of our action.

Jean-Marie Robine
James Vaupel
Bernard Jeune
Michel Allard

Acknowledgements: The editors wish to thank Mary Lynn Gage and Isabelle Romieu for editorial assistance and Jacqueline Mervaillie for the organization of the meeting in Paris.

Contents

Contributors

Allard, M.
Fondation IPSEN, 24, Rue Erlanger, 75781 Paris, Cedex 16, France

Caselli, G.
Università La Sapienza, Dipartimento di Scienze Demografiche,
Via Nomentana 41, 00161 Roma, Italy

Duhon, S.A.
Dept. Psychology, Inst. Behav. Genetics, University of Colorado, Boulder,
CO 80309-0447, USA

Finch, C.
Division of Neurogerontology, Andrus Gerontology Center,
University of Southern California, Los Angeles, CA 90089-0191, USA

Forette, B.
Hôpital Sainte Périne, 11, Rue Chardon-Lagache, 75016 Paris, France

Jeune, B.
Odense University, Center for Health and Social Policy, Winslowparken 17,
DK 5000 Odense, Denmark

Johnson, T.E.
Dept. Psychology, Inst. Behav. Genetics, University of Colorado, Boulder,
CO 80309-0447, USA

Johnson, M.A.
Food and Nutrition, University of Georgia, Dawson Hall, Athens, GA 30602-1775,
USA

Kannisto, V.
Campo Grande 1,6-D, 1700 Lisbon, Portugal

Kirkwood, T.
Biological Gerontology Group, University of Manchester, 3239 Stopford Buil.
5655 Oxford Rd, Manchester, M13 9 PT, U.K.

Martin, P.
Human Development & Family Studies, Iowa State University, 209, CD Building, Ames, IA, 50011, USA

Mockett, R.J.
Department of Biological Sciences, Southern Methodist University, Dallas, TX 75275, USA

Murakami, S.
Dept. Psychology, Inst. Behav. Genetics, University of Colorado, Boulder, CO 80309-0447, USA

Olshansky, S.J.
Center on Aging, Health and Society, University of Chicago, 5841 S. Maryland Avenue, MC 6098, Chicago, IL 60439, USA

Poon, L.
Gerontology Center, University of Georgia, 100 Candler Hall, Athens, GA 30602-1775, USA

Robine, J.-M.
Equipe Démographie et Santé, Centre Val d'Aurelle – Parc Euromédecine, 326, Rue des Apothicaires, 34298 Montpellier Cedex 5, France

Schächter, F.
Centre Etudes Polymorphisme Humain, 27, Rue Juliette Dodu, 75010 Paris, France

Shook, D.R.
Dept. Psychology, Inst. Behav. Genetics, University of Colorado, Boulder, CO 80309-0447, USA

Sohal, R.S.
Department of Biological Sciences, Southern Methodist University, Dallas, TX 75275, USA

Tedesco, P.M.
Dept. Psychology, Inst. Behav. Genetics, University of Colorado, Boulder, CO 80309-0447, USA

Vallin, J.
Centre français sur la Population et le Développement (CEPED), 15, rue de l'Ecole de Médecine, 75270 Paris Cedex 06, France

Vaupel, J.W.
Odense University, Center for Health and Social Policy, Winslowparken 17,
5000 Odense, Denmark

Yi, Zeng
Institute of Population Research, Peking University, Beijing 100871, China

Zhenglian, W.
Institute of Population Research, Peking University, Beijing 100871, China

Practical Limits to Life Expectancy in France

S. J. Olshansky

Introduction

The expected duration of life for a baby born in France in 1991 was 76.9 years –
72.9 years for males and 81.2 years for females (Couet and Tambay 1995). These
figure are calculated from period life tables produced from age-specific death
rates that prevailed in France from 1990–1992. Estimates of the duration of life
based on period life tables are predicated on the assumption that babies born in
a given year will experience, for the duration of their lives, the prevailing mortality risks observed at every age. When secular declines in total mortality are
occurring, as they have been in France and other developed nations for most of
the 20th century, estimates of life expectancy at birth using this method tend to
underestimate the subsequent longevity of the birth cohort. Given the historical
trend in declining death rates in France and other developed nations, and the
expectation that they will continue to decline in the future, how much higher can
life expectancy at birth be expected to rise beyond these period estimates? Is the
measure of period life expectancy providing a reasonable estimate of future longevity, or is it possible that the actual life expectancy of babies born in France
today will be considerably higher than is currently indicated by period life tables?
Alternatively, how much higher can life expectancy at birth practically increase
in France and other developed nations with life expectancy (for males and females combined) approaching 80 years?

Previous efforts to estimate how high life expectancy can rise have been
based almost exclusively on attempts to answer a simple question: how much
lower can death rates decline? French demographer Bourgeois-Pichat (1952,
1978) addressed this question by partitioning total mortality into its extrinsic
and intrinsic (e.g. biological) elements and then estimated the lower bounds to
intrinsic mortality. Fries (1980) based his estimate of the upper limits to human
longevity on the extrapolation of past mortality trends into the future – a method
still used by United States government actuaries for forecasting trends in population growth and aging (Bell et al. 1993; Day 1993). Similar methods of estimating
how low death rates can decline have been used by others (Ahlburg and Vaupel,
1990; Manton, et al. 1991; Siegel 1980).

An alternative approach to addressing this question is to estimate the age-
specific mortality reductions required to produce life expectancies that are
higher than those prevailing today. That is, instead of asking how low death rates

J.-M. Robine et al. (Eds.)
Longevity: To the Limits and Beyond
© Springer-Verlag Berlin Heidelberg New York 1997

can decline, a "reverse engineering" approach leads to the question, how low would death rates have to decline in order for life expectancy to rise from its current level to some specified higher number? The mortality schedules and patterns of survival for populations derived by this method may then be evaluated with respect to their plausibility.

Previous research using the reverse engineering approach to estimating practical limits to life expectancy has determined that extremely large reductions in death rates at every age would be required in the United States to increase life expectancy at birth (for males and females combined) beyond 85 years (Olshansky et al. 1990). For example, a life expectancy at birth of 85 years requires that death rates decline by 65 % at every age from levels observed in 1985 (Fig. 2 in Olshansky et al. 1990). Since the elimination of all deaths attributable to heart disease and cancer would not reduce death rates enough to achieve a life expectancy of 85 years, it was viewed as impractical to expect life expectancy to rise beyond this level without significant new advances in medical technology.

An evaluation of recent and historical patterns of mortality in the United States indicates that the rapid declines in death rates required to produce a life expectancy that exceeds 85 years at any time in the 21st century are not occurring (Olshansky and Carnes 1994). In fact, the latest mortality statistics for the United States indicate that life expectancy at birth actually declined by 0.3 years (from 75.8 to 75.5) from 1992 to 1993 (Gardner and Hudson 1996). This decline occurred principally because of a rise in death rates from heart disease, COPD, HIV infection, accidents, pneumonia and influenza, diabetes, and stroke. The increasing difficulty in pushing life expectancy higher in low mortality populations has been referred to as entropy in the life table, a concept that has been well documented in the scientific literature (Horiuchi 1989; Keyfitz 1977, 1985; Olshansky et al. 1990; Rogers 1995).

If entropy in the life table is characteristic of human populations with life expectancies at birth that approach 80 years, then this phenomenon should be observed in France, where average life expectancy has already reached 77 years. In this paper the phenomenon of entropy in the life table is demonstrated using mortality data for the French population between 1990 and 1992.

Data and Methods

The source of data for this study is a complete schedule of conditional probabilities of death $[q_{(x)}]$ by sex and single-year-of-age (from ages 0–99) for France from 1990 to 1992 (Couet and Tambay 1995). The complete mortality schedules required to achieve higher life expectancies (from observed levels to 90 years) were estimated by reducing death rates proportionally at every age, or restricting the proportional reductions to those aged 50 and older. Estimates of survival to ages 65, 85, and 100 were based on proportional reductions in death rates at every age required to produce life expectancies at birth up to 95 years. The proportional reductions were iterated until each of the targeted life expectancies was achieved. The life table was closed at the age of 100.

Closing the life table at 100 can lead to a problem when a large proportion of a birth cohort is projected to survive beyond this age range. The potential problem is that, because all survivors to the age at which the life table is closed are projected to die at that age, life expectancy will be underestimated, because some people who survive to the age of 100 will live some months or years beyond that age. However, the underestimation of life expectancy at birth resulting from this problem has been shown to be insignificant until after the projected life expectancy exceeds 90 years (Olshansky and Carnes 1996). It is for this reason that the projected life expectancies in this study do not exceed the age of 90. Although the proportion of the birth cohort expected to survive to selected older ages is based on a projected life expectancy that reaches 95, the interpretation of the survival $[l_{(x)}]$ column of the life table is unaffected by closing the life table at that age. Closing the life table at an age where few or no people survive (e.g., beyond 110–120 years) requires assumptions about anticipated patters of survival rarely experienced by humans. A study of this sort has already been performed (e.g., see Olshansky and Carnes 1996).

Results

Mortality Reductions for Higher Life Expectancies

As demonstrated previously for the United States (Olshansky et al. 1990), extremely large reductions in death rates from prevailing levels are required at every age to produce significant increases in life expectancy at birth in France. In order for life expectancy in France (males and females combined) to rise from its current level of 76.9 years to 80 years, death rates at every age would have to decline by 23 % (Fig. 1). A 52 % reduction in death rates at every age would be required to increase life expectancy to 85 years, and a 74 % reduction in death rates at every age would be required to increase life expectancy to 90 years. Female life expectancy at birth in France was 81.2 years in 1990–92. An additional 4.8 years in life expectancy (i.e. with $e_{(o)} = 85$) would require a 33.2 % reduction in death rates at every age while a life expectancy at birth of 90 years would require a 64.1 % reduction in death rates at every age.

Since deaths rates at younger ages (before age 50) are already very low among females in France, rapid gains in longevity in the future must ultimately result from reductions in death rates after age 50. If mortality reductions in the future are restricted to the 50 and older age range, the magnitude of the reductions in death rates required to increase female life expectancy to 85 and 90 years would be higher (e.g., 39.7 % instead of 33.2 % with $e_{(o)} = 85$, and 73.1 % instead of 64.1 % with $e_{(o)} = 90$; Fig. 1). Comparable differences in death rates occur for males.

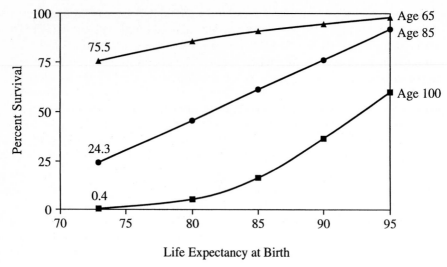

Fig. 1. Percentage of reduction in the conditional probability of death [q_x] required to increase life expectancy at birth in France from current levels to between 80 and 90 years

Cohort Survival

An alternative method of gauging the plausibility of achieving much higher life expectancies is to examine required changes in patterns of survival. Based on period life tables for France in 1990–92 in which life expectancy at birth for

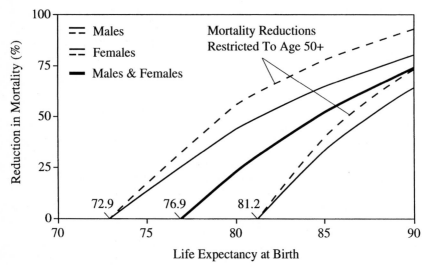

Fig. 2. Percentage of 1990–92 male birth cohort in France required to survive to ages 65, 85, and 100 for life expectancy at birth to rise to 75 to 95 years

females was 81.2 years, 89.4 % of the original birth cohort are expected to survive to the age of 65, 48.5 % are expected to survive to their 85th birthday, and 1.7 % are expected to become centenarians (Fig. 2). A life expectancy of 85 years would require 92.8 % survival to age 65, 61.9 % survival to age 85, and 7.2 % survival to the age of 100.

Once the projected life expectancy at birth for females in France exceeds 85 years and approaches 90 or 95 years, the patterns of survival required to achieve higher life expectancies change dramatically. For example a life expectancy at birth of 95 years requires 98.5 % survival to age 65, 90.7 % survival to age 85, and 59.9 % survival to the age of 100. Similar patterns of survival are required for males to achieve gains in longevity that are comparable to females. Because the required patterns of survival change in an even more dramatic fashion when life expectancy at birth reaches 100 years (Olshansky et al. 1990; Olshansky and Carnes 1996), such a scenario is considered highly implausible.

In order for life expectancy at birth to rise to 95 years or higher, all intrinsic mortality[1] would have to be eliminated before age 65 (including inherited lethal genetic disorders), and significant declines in death rates from extrinsic mortality (principally accidents, homicide, and suicide) would be required. This is the case because intrinsic mortality is already very low in France at younger and middle ages. In fact, a life expectancy of 95 for males in France would require an increase in survival to the age of 100 from the less than one-half of one percent under prevailing mortality conditions, to 60 percent survival – a 150-fold increase. At this time there is no reason to expect that these extremely favorable survival conditions can be achieved in France or any other developed nation. In fact, it is not currently possible to eliminate inherited lethal genes at any age, neither is there justification at this time to support the assumption that intrinsic causes of death can be eliminated or even dramatically reduced.

It should be emphasized in this study that estimates of higher life expectancies are based on proportional reductions in death rates either throughout the entire age range or only among those aged 50 and older. It is theoretically possible to achieve a life expectancy at birth of 95 years or higher without having pro-

---

[1] Intrinsic mortality is defined as causes of death that result from 1) the expression of inherited lethal genes at any age in the life span, 2) the expression of disease processes arising from endogenously acquired genetic damage (e.g., metabolic free radicals), 3) traditional senescent-related diseases and disorders arising from the progressive deterioration of cells, tissues, organs, and organ systems (resulting from some combination of inherited and acquired damage), and 4) intrinsic causes of death that have been influenced, either positively of negatively, by lifestyle modification, living conditions, or medical technology (for a more detailed discussion of this issue see Carnes et al., 1996). Intrinsic mortality differs from senescent mortality, in that deaths are anticipated throughout the age structure. Under this partitioning of total mortality, intrinsic mortality is a subset of total mortality, an senescent mortality is subset of intrinsic mortality. A more precise enumeration of intrinsic causes of death will also emerge as biomedical researchers improve their understanding of the genetic mechanism that are either responsible for or closely linked to such causes of death as cancer and heart disease (Carnes and Olshansky 1997). This definition of intrinsic mortality acknowledges that humans have developed an ability to influence intrinsic disease processes.

portional reductions in death rates and 60 % of the original birth cohort surviving to become centenarians. However, in order for this to occur, almost everyone in a given birth cohort would have to survive to, and die within, an extremely narrow age range near the projected life expectancy at birth – a condition referred to as a square survival curve (Fries 1980). Yet historical trends in death rates are not leading to a squaring of the survival curve (e.g.; see Myers and Manton 1984). It has also been suggested that much higher life expectancies can be achieved by delaying intrinsic mortality rather than eliminating it. Delaying mortality from intrinsic causes has the same effect on age-specific death rates and the mortality schedule as eliminating individual causes of death; in either case, period age-specific death rates at younger ages would have to be reduced to near zero in order for life expectancy at birth to reach 90 years or higher. Is this practically achievable in a genetically heterogeneous population composed of some individuals who carry inherited lethal genes, and in a world where extrinsic sources of mortality appear to be ubiquitous and unavoidable?

Comparing Period and Cohort Life Table Estimates

Finally, the practical problem of increasing life expectancy further in low mortality populations can also be illustrated by comparing longevity estimates from period and cohort life tables. The expected duration of life for babies born in France in 1900 based on a period life table was 43.4 years for males and 47.0 years for females (Fig. 3). A cohort life table for this population, which is based on the observed pattern of survival throughout the 20th century for babies born in 1900, demonstrates that their observed longevity was actually 48.1 years and 55.4 years, respectively. This means that, during a time when death rates declined dramati-

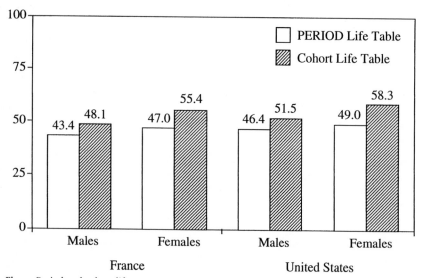

Fig. 3. Period and cohort life expectancy at birth in France and the United States in 1900, by sex

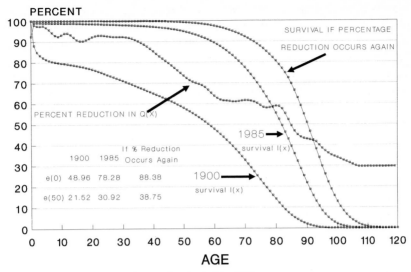

Fig. 4. Percentage of reduction in q_x for U.S. females (1900–1985)

cally throughout the age structure in France, including large declines in early age mortality at the beginning of the century and rapid declines at middle and older ages in the last third of the century, the use of a period life table resulted in an underestimate of life expectancy at birth of 4.7 years for males and 8.4 years for females. For the United States, the 1900 period life table underestimated cohort longevity by 5.1 years for males and 9.3 years for females.

Even if the magnitude of the reduction in death rates at every age occurred again in France or in the United States from levels prevailing today, the underestimation of life expectancy from a period life table would be much smaller than it was in 1900 (Olshansky et al. 1991). For example, the large percentage declines in death rates observed at every age in the United States from 1900 to 1985 produced a 29.3 year increase in life expectancy at birth. If comparable percentage reductions occurred again at every age from current levels of mortality, the gain in life expectancy at birth would only be 10.1 years (Fig. 4). This phenomenon occurs when comparable percentage declines in death rates are applied in increasingly more favorable mortality schedules (i.e., lower death rates). The result is a decline in absolute reductions in death rates.

Conclusions

The rapid increase in life expectancy throughout most of the 20th century in France and other developed nations is attributable to reductions in death rates at younger ages, primarily from infectious and parasitic diseases and maternal mortality. Since death rates before age 50 have been reduced dramatically, it is now common in low mortality populations to have 90 percent or more of every birth cohort survive past that age. In fact, patterns of survival have improved so dramatically in France that up to 50 % of the current birth cohort of females can expect to survive to at least their 85th birthday. This means that if rapid gains in life expectancy are to be achieved in the future, they will have to result from mortality reductions at older ages. Although relatively rapid declines in death rates occurred at older ages in the past, rapid gains in longevity in the future can only be realized if the remaining causes of death (which are much lower now than they were just 25 years ago) decline precipitously.

In this study it was demonstrated that in order for life expectancy at birth in France to increase from its current level of 76.9 years to 85 years, mortality at every age would have to be reduced by 52 %. Since mortality at younger and middle ages is already extremely low in France, larger reductions in death rate would be required at older ages in order for this higher life expectancy to be achieved. In addition, such declines would have to be concentrated among the major fatal diseases – implying that much larger cause-specific reductions in death rates are required to reduce total mortality by half. Are such large reductions in cause-specific death rates practically achievable? It has been suggested that large gains in longevity are practically achievable by modifying lifestyles at the population level (Manton et al. 1991).

Although many studies have focused on the health benefits associated with lifestyle modification (e.g., see Anderson et al. 1987; Frick et al. 1987; Lipid Research Clinic Program 1984; Muldoon et al. 1990; Paffenbarger et al. 1986, 1993), there is no empirical evidence suggesting that any single intervention or lifestyle modification, or combination of interventions, would lead to the mortality reductions required to reduce total mortality by half at every age. In fact, clinical trials and retrospective health studies suggest that mortality would not change much at all, even if the entire population adopted optimum risk factor profiles (Epstein 1992; Muldoon et al. 1990; Rogers 1995). It is therefore improbable that France will achieve a life expectancy at birth that exceeds 85 years (for males and females combined) in the near future given what is currently known about lifestyle modification and medical technology.

It is important to recognize that the argument developed thus far does not imply that life expectancy at birth cannot rise to 85 years or higher. Indeed, there is a considerable amount of promising research in the fields of molecular biology, genetics, and other disciplines suggesting that the basic rate of aging itself may eventually fall, to some extent, within the control of medical technology. Life expectancy at birth can rise beyond 85 years, but it is suggested here that this would require significant new advances in medical technology that "manufacture survival time" by decelerating the basis rate of aging itself and postponing death through medical interventions. However, given what is known today about the inescapable mathematics of entropy in the life table and the limited gains in longevity associated with lifestyle modification and medical technology, the practical limits to life expectancy at birth in France and other developed nations are being rapidly approached. Because large reductions in death rates (e.g., 52 % at every age) are required to raise life expectancy at birth in France by an additional

8.1 years (to 85 years), the suggestion that a life expectancy of 85 years is practically achievable may, in itself, be an optimistic scenario.

References

Ahlburg DA, Vaupel JW (1990) Alternative projections of the U.S. population. Demography 27(4):639–652

Anderson KM, Castelli WP, Levy D (1987) Cholesterol and mortality: 30 years of follow-up from the Framingham study. J Am Med Assoc 275(16):2176–2180

Bell FC, Wade AH, Goss SC (1993) Life tables for the United States Social Security Area 1900–2080. Actuarial study no. 107, Social Security Administration, SSA Publication No. 11–11536

Bourgeois-Pichat J (1952) Essai sur la mortalité biologique de l'homme. Population 7(3):381–394

Bourgeois-Pichat J (1978) Future outlook for mortality decline in the world. Population Bull UN11:12–41

Carnes BA, Olshansky, SJ (1997) A biologically motivated partitioning of mortality. Experimental gerontology (in press)

Carnes BA, Olshansky SJ, Grahn D (1996) Continuing the search for a law of mortality. Population Dev Rev 22(2):231–264

Couet C, Tambay L (1995) "La situation démographique en 1992" Insee Résultats No. 386–387

Day JC (1993) Population projections of the United States, by age, sex, race, and Hispanic origin: 1992–2050. Curr Pop Rep, Series P-25, No 1092

Epstein FH (1992) Low serum cholesterol, cancer and other noncardiovascular disorders. Atherosclerosis 94:1–11

Frick MH, Elo O, Happa K, Heinonen OP, Heinsalmi P, Helo P, Huttunen JK, Kaitaniemi P, Koskinen P, Manninen V (1987) Helsinki heart study: Primary-prevention trail with Gemfibrozil in middle-aged men with dyslipidemia. New Engl J Med 316:1237–1245

Fries JF (1980) Aging, natural death, and the compression of morbidity. New Engl J Med 305:130–135

Gardner P, Hudson BL (1996) Advance report of final mortality statistics. 1993 Monthly Vital Stat Rep 44(7-S):1–13

Horiuchi S (1989) Some methodological issues in the assessment of the deceleration of the mortality decline. In: Differential mortality: methodological issues and biosocial factors. Ruzika L, Wunsch G, Kane P (eds) Oxford, Clarendon Press, pp 64–78

Keyfitz N (1977) What difference would it make if cancer were eradicated? An examination of the Taeuber paradox. Demography 14:411–418

Keyfitz N (1985) Applied mathematical demography. 2nd ed. New York, Springer.

Lipid Research Clinic Program (1984) Coronary primary prevention trial results: I. Reduction of incidence of coronary heart disease. J Am Med Assoc 251:351–364

Manton KG, Stallard E, Tolley HD (1991) Limits to human life expectancy. Pop Devel Rev 17(4):603–637

Muldoon MF, Manuck SB, Matthews KA (1990) Lowering cholesterol concentrations and mortality: A quantitative review of primary prevention trials. Br Med J 301:309–314

Myers G, Manton K (1984)Compression of mortality: myth or reality? Gerontologist 24–346

Olshansky SJ, Carnes BA (1994) Demographic perspectives on human senescence. Pop Devel Rev 20(1):57–80

Olshansky SJ, Carnes BA (1996) Prospect for extendend survival: a critical review of the biological evidence. In: Health and mortality among elderly populations. Caselli G, Lopez A (eds) Clarendon Press, Oxford: 39–58

Olshansky SJ, Carnes BA, Cassel C (1990) In search of Methuselah: estimating the upper limits to human longevity. Science 250:634–640

Olshansky SJ, Rudberg MA, Carnes BA, Cassel C, Brody J (1991) Trading of longer life for worsening health: The expansion of morbidity hypothesis. J Aging Health 3(2):194–216

Paffenbarger R, Hyde RT, Wing AL, Lee IM, Jung DL, Kampert JB (1993)The association of changes in physical-activity level and other lifestyle characteristics with mortality among men. New Engl J Med 328(8):538–545

Paffenbarger RS, Hyde RT, Wing AL, Hsieh C (1986) Physical activity all-cause mortality, and longevity of college alumni. New Engl J Med 314:605–613

Rogers RG (1995) Sociodemographic characteristics of long lived and healthy individuals. Pop Devel Rev 21(1):33–58

Siegel JS (1980) NIH Publ 80–969, pp 17–82

The Average French Baby May Live 95 or 100 Years

J. W. Vaupel[*]

If death rates at each age in France remained at current levels over the lifetimes of babies born in France this year, then more than half the babies would live to celebrate their 80th birthdays. Among baby girls, two-thirds would become octogenarians and half would reach age 85. Death rates have been declining in France (and in most other developed countries as well) for many decades. In particular, death rates among octogenarians and nonagenarians have fallen substantially since 1950. Extrapolating these rates of improvement into the future yields an astonishing result: half of all French babies may survive to celebrate their 95th birthdays and half of French girl babies may become centenarians.

Whether progress in reducing mortality will continue, decelerate, or accelerate is an open question. Biomedical research may fail to continue to produce the advances needed to save lives. Social and economic conditions may become unfavorable. Environmental conditions may substantially deteriorate. Nuclear war may kill millions or even billions. New diseases, like AIDS, may decimate populations.

On the other hand, biological, medical and gerontological breakthroughs could lead to considerable extensions of the human life span. The life sciences may be poised at roughly the point the physical sciences were a century ago: biological innovations comparable to electricity, automobiles, telephones, television, rockets, and computers may be forthcoming. Fundamental advances could occur over the coming decades in genetic engineering, in the prevention and treatment of such diseases as arteriosclerosis, cancer, diabetes, and dementia, and perhaps even in understanding and controlling human aging itself. It will be 80 years before a newborn turns 80; a great many unanticipated advances may be made over those eight decades.

The future is not just uncertain, it is surprisingly uncertain. As Ascher (1978), Keyfitz (1981), Stoto (1983) and others reviewed by Ahlburg and Land (1992) have demonstrated, the actual course of demographic events often leads to outcomes beyond the most extreme projections. Consequently, forecasts ought to include wide bands of uncertainty that spread outward at an expanding pace into the more and more distant future. On the one hand, the year 2096 is less than a life span away: some children alive today will almost certainly still be alive then.

---

[*] Odense University Medical School, Odense, Denmark

J.-M. Robine et al. (Eds.)
Longevity: To the Limits and Beyond
© Springer-Verlag Berlin Heidelberg New York 1997

On the other hand, the most radical changes can occur in a lifetime; the year 2096 may be as astonishingly different from 1996 as 1996 is from 1896.

Under the mortality regime that prevailed in France in 1896, only three new-born boys and four newborn girls in 100 would live to age 85. Under the mortality regime that prevails in France today, a quarter of newborn boys and half of newborn girls will live to 85. Some of this change reflects improvements in infant, child, and young adult mortality, but a substantial proportion reflects progress in reducing mortality at older ages. At 1896 death rates, only one girl in two could expect to live to age 60. Of the women who reached age 60, less than one in eight would reach age 85. At current death rates, over 95 % of baby girls can expect to reach age 60, so survival to 60 has doubled, from 50 % to 95 %. At current death rates, half of the women who reach age 60 can expect to survive to age 85, so survival from 60 to 85 has quadrupled, from less than one in eight to one in two. This radical improvement in life chances, especially at older ages, could hardly have been foreseen in 1896.

If something goes badly wrong – environmental collapse, nuclear devastation, etc. – then the cohort born in 1996 may lead short, miserable lives. If health, economic, and social progress accelerate, then there is some chance that most of the members of this cohort will survive into the 22nd century. My best guess among this very wide range of possibilities is that half of the babies born in France this year will reach age 95 and that most of the girls will reach age 100. This is speculation, as are all predictions, and perhaps more tantalizing than informative. What definitely is informative, however, and what definitely constitutes an important intellectual advance is the new light that has been cast on longevity and survival by new data sets. My speculation about the life spans of French babies is informed speculation based on paradigm-shattering findings from these new data sets.

In particular, until recently little was known about the plasticity of oldest-old mortality. Demographers conjectured that mortality at advanced ages was intractable. Specifically, they assumed that rates of mortality improvement at older ages were slow, decelerated with increasing age, and decelerated with increasing life expectancy. These assumptions were justified by appeal to three related notions:

1) Deaths at older ages are essentially due to old age, and nothing can be done about old age;
2) The typical human organism is not constructed to survive much past age 80 or 90;
3) Causes of death at younger ages are largely extrinsic but causes of death at older ages are mostly instrinsic, and it is very difficult to reduce intrinsic causes of death.

These beliefs led most demographers to predict that human life expectancy would not increase very much in the future.

Evidence is now available about mortality after age 80 in a variety of developed countries for several decades. Biological evidence is also now available that permits deeper understanding of the genetic and environmental factors affecting longevity and mortality in a variety of species. This new evidence is inconsistent with previously accepted views. In particular, mortality improvements at advanced ages have been large and have accelerated as life expectancy has increased. Depending on the country examined and the analytical perspective adopted, mortality improvements may not decline with age. In addition, the new evidence casts doubt on the notions that nothing can be done about old age, that organisms are not constructed to survive past some age, and that it is useful to distinguish between extrinsic and intrinsic causes of death. It is this new evidence and its implications that are the focus of this essay.

Prior to 1990, only a few, scattered pieces of research shed light on mortality at the oldest-old ages above 80. Key contributions included Vincent (1951), Depoid (1973), Thatcher (1987), and Kannisto (1988). In particular, very little was known about the plasticity of oldest-old mortality. Lacking empirical studies of large numbers of life tables for the oldest ages over time and place, demographers surmised that mortality at advanced ages was intractable. Specifically, they assumed that rates of mortality improvement at older ages were slow, decelerated with increasing age, and decelerated with increasing life expectancy. In estimating model life tables at low levels of mortality, Coale and Guo (1989) relied on all three of these assumptions. So did Bourgeois-Pichat (1952, 1978) and Demeny (1984) in making forecasts about upper limits to human life expectancy, forecasts that already have been exceeded in some countries. As Keilman (1995) has documented, for several decades nearly all the forecasts produced by national statistical offices have erroneously assumed that rates of mortality improvement at older ages would be slow and would decline with time.

Because of painstaking work by a group of diligent demographers, death counts and population counts by single year of age up to the highest ages and by single year of time back four decades or more are now available in the Odense Archive of Population Data on Aging for most developed countries, including France. These data were largely compiled and organized by Väinö Kannisto, supplemented by data for England and Wales provided by A. Roger Thatcher, data for Sweden from Hans Lundström, data for Norway from Jens Borgan, and data for Denmark from Kirill Andreev and Axel Skytthe. Conjecture can now be replaced by demographic analysis. A stream of research based on the new data is beginning to appear, including Kannisto (1994, 1996), Kannisto et al. (1994), Vaupel and Lundström (1994), Manton and Vaupel (1995), Vaupel and Jeune (1995a), and Thatcher et al. (1997).

This essay adumbrates the major findings reported in this earlier research and adds further findings contained in new tables and figures. The basic thrust of the research is that mortality at older ages is plastic and has been substantially reduced in recent decades. Rates of improvement have tended to accelerate with time.

Table 1. Central death rates for aggregate of Denmark, Finland, Norway, and Sweden, for males and females, for sexagenarians, septuagenarians, octogenarians, nonagenarians, and centenarians in 1880–99, 1930–49, and 1989–93, as well as ratio of values in 1880–99 to values in 1989–93, ratio of values in 1930–49 to values in 1989–93, and annual average rates of mortality improvement between 1880–99 and 1989–93 and between 1930–49 and 1989–93.[1]

Sex	Age category	Central death rates (%)			Difference between rate in this period and rate in 1989–93		Ratio of rate in this period to rate in 1989–93		Annual average rate of mortality improvement from this period until 1989–93	
		1880–99	1930–49	1989–93	1880–99	1930–49	1880–99	1930–49	1880–99	1930–49
Males	60–69	3.7	3.0	2.2	1.5	0.8	1.68	1.38	0.51	0.62
	70–79	8.5	7.2	5.5	3.0	1.7	1.54	1.30	0.43	0.51
	80–89	19.6	17.3	13.3	6.3	4.0	1.47	1.30	0.38	0.51
	90–99	38.4	36.5	28.7	9.7	7.8	1.34	1.27	0.29	0.47
	100+	95.9	76.9	52.9	43.0	24.0	1.81	1.46	0.59	0.73
Females	60–69	3.1	2.4	1.1	2.0	1.3	2.72	2.12	0.98	1.44
	70–79	7.5	6.4	3.1	4.4	3.3	2.45	2.11	0.88	1.44
	80–89	17.2	16.1	9.1	8.1	7.0	1.89	1.77	0.62	1.10
	90–99	34.5	33.9	23.4	11.1	10.5	1.47	1.45	0.38	0.71
	100+	66.3	70.1	48.5	17.8	21.6	1.37	1.46	0.31	0.73

[1] Statistics in this and other tables and figures in this essay were calculated as follows. The annual age-specific central death rate is given by

$$m(x,y) = D(x,y) / [N(x,y)+N(x,y+1))/2],$$

where $D(x,y)$ represents the number of deaths at age x over the course of year y among males or females, and $N(x,y)$ represents the number of males or females who were x years old on January 1 of year y. The average death rate in the interval from age x through age x^* and year y through year y^* can be calculated by

$$\bar{m}(x,x^*, y,y^*) = \left[\sum_{j=y}^{y^*} \sum_{i=x}^{x^*} w(i)\, m(i,j) \right] / \left[\sum_{j=y}^{y^*} \sum_{j=y}^{x^*} w(i) \right].$$

The weights w are used to standardize the sex and age composition of the population so that comparisons can be made over time, across populations, and between sexes. We based the weights on the age composition of the elderly Swedish population, males and females combined, from 1960 through 1993:

$$w(i) = \left[\sum_{y=1960}^{1993} N_m(i,y)+N_f(i,y) \right] / \left[\sum_{x=50}^{111} \sum_{y=1960}^{1993} N_m(x,y)+N_f(x,y) \right],$$

where N_m and N_f denote male and female population counts. Sometimes it was impossible to estimate m for a specific year either because no one was alive at that age and year or because we did not have data for that age and year. In such cases, the m term was dropped from the numerator and the corresponding weight was dropped from the denominator of the expression for $\bar{m}$. All death rates reported in this essay are values for $\bar{m}$. The average annual rate of improvement in mortality from the first period to the second period is given by

$$\varrho = 1-(\bar{m}_2 / \bar{m}_1)^{1/\delta},$$

where $\bar{m}$ is defined in the note to Table 1 and where δ is the interval between the means of the two periods

$$\delta = (y_2+y_2^*)/2 - (y_1+y_1^*)/2,$$

the first period running from y_1 through y_1^* and the second from y_2 through y_2^*.

Findings

The plasticity of oldest-old mortality is clearly revealed in historical perspective. The four major Nordic countries – Denmark, Finland, Norway, and Sweden – have reliable data on population counts and death counts up to the highest ages going back more than a century. In Table 1 central death rates are presented for sexagenarians, septuagenarians, octogenarians, nonagenarians, and centenarians in the aggregate of these four countries combined as if they constituted a single country. The death rates pertain to three time periods: 1880–1899, 1930–1949 and 1989–1993. The dramatic decline in death rates at older ages is apparent from comparison of the death rates in the three periods.

Table 1 provides three measures of mortality decline. The first measure is the absolute difference between death rates in the earlier periods and the most recent periods. Although demographers seldom use this measure, it is a reasonable way of indicating how many deaths were averted and lives saved. Remarkably, the older the age category, the greater was the absolute reduction in death rates. Viewed in this way, mortality becomes increasingly plastic with age. Except for centenarians, greater progress was made for females than for males. The centenarian statistics have to be viewed with caution because there were few centenarians until recently (Vaupel and Jeune 1995b). In the calculations for 1880–1899, the number of male centenarian person-years was less than 100.

It might be objected that the lives of the extremely old are not "saved" for very long. The number of deaths averted at the oldest ages was an order of magnitude greater than the number of deaths averted among the younger elderly, but the remaining life expectancy of the younger elderly is an order of magnitude greater. Furthermore, the quality of life of some very old persons may tend to be, along some dimensions, lower than the quality of life at younger ages. Nonetheless, averting deaths is an important medical and public-health achievement, as well as generally being of inestimable value to the persons whose lives are saved. The fact that the absolute decline in mortality is greatest at the oldest ages is thus of considerable significance.

The second measure presented in Table 1 is the ratio of death rates in the earlier periods relative to the most recent period. The third measure is the annual average rate of mortality improvement between the earlier periods and the recent period. (How these measures were calculated is explained in the note to the table). These two measures are closely related and show the same patterns. Female mortality improved more than male mortality. For females there is a marked decline in the rate of improvement with age, but for males such a decline is much less apparent. For both males and females the rate of mortality improvement accelerated, being substantially greater over the last 50 or 60 years than it was over the last century.

These various findings from Table 1 are consistent with results from an analysis of mortality in Sweden since 1900 (Vaupel and Lundström 1994) and are in

Table 2. Average annual rates of improvement in mortality (in percent) for aggregation of Denmark, Finland, Norway and Sweden, for males and females, for sexagenarians, septuagenarians, octogenarians, and nonagenarians, over successive periods 20 years apart.[1]

Sex	Age category	Time period				
		1880–9 to 1900–9	1900–9 to 1920–9	1920–9 to 1940–9	1940–9 to 1960–9	1960–9 to 1980–9
Males	60–69	0.55	0.36	0.45	0.20	0.60
	70–79	0.48	0.28	0.31	0.21	0.50
	80–89	0.26	0.26	0.22	0.32	0.64
	90–99	−0.02	0.12	0.13	0.36	0.75
Females	60–69	0.76	0.32	0.68	1.71	1.54
	70–79	0.41	0.18	0.40	1.04	2.05
	80–89	0.16	0.09	0.20	0.63	1.74
	90–99	−0.01	0.02	0.02	0.50	1.18

[1] See note to Table 1 for calculation details

general agreement with the patterns since 1950 in 28 developed countries uncovered by Kannisto et al. (1994) and Kannisto (1994, 1996).

Further information about the pattern of mortality decline in the Nordic countries is provided in Table 2. The more rapid rate of improvement for females than males can be seen to be a persistent phenomenon, although the female advantage has tended to widen over time. The rate of improvement accelerated substantially, especially at the older ages and in the most recent periods, and the gain for females was even more striking than the gain for males. For females, the rate of improvement generally declines with age, although in the most recent period the rate of improvement was greater for septuagenarians and octogenarians than for sexagenarians. For males, the rate of improvement declined with age before the 1940s but afterwards the rate of improvement for octogenarians and nonagenarians has been greater than among the younger elderly.

For 13 countries in the Kannisto-Thatcher Oldest-Old Database, reliable population and death counts are available from 1960 to 1992. These countries are Austria, Belgium, Denmark, England (including Wales), Finland, France, the former West Germany, Japan, the Netherlands, Norway, Scotland, Sweden and Switzerland. Figure 1 graphs average annual improvements in mortality for these countries aggregated as if they were a single country. The improvements are measured as average annual changes over successive five-year-periods; the last data point, for instance, pertains to the improvement between 1983–1987 and 1988–1992. A striking feature of the figure is the sharp increase in rates of improvement around 1970: for both males and females and for octogenarians as well as nonagenarians, rates of improvement approximately doubled. Since this transition, however, rates of improvement have hovered up and down around a roughly constant level. Also noteworthy in the figure is the large difference between rates of improvement for female octogenarians compared with the other three groups.

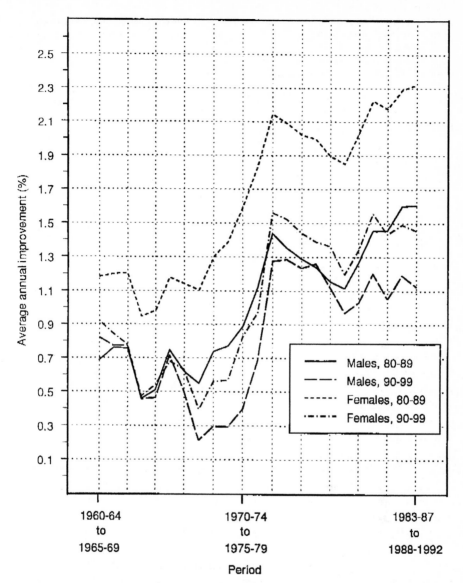

Fig. 1. Average annual improvement in mortality between successive five-year periods, for male and female octogenarians and nonagenarians, in an aggregate of 13 countries

Kannisto et al. (1994) present a similar graph, but for nine countries up to 1989 and for successive ten-year as opposed to five-year periods. Trends in rates of mortality improvement are so important that further analysis is required on a country-by-country basis, using various time periods and using new data for more recent years as they becomes available.

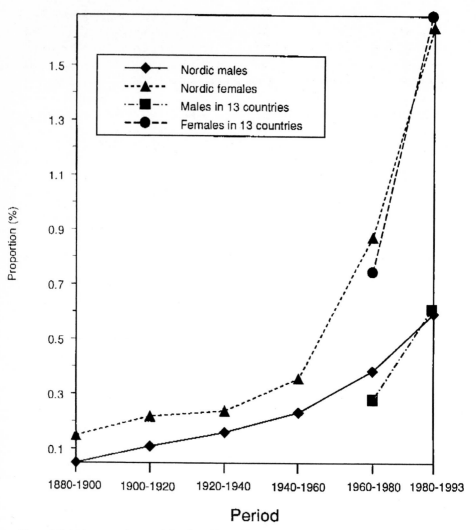

Fig. 2. Life table proportion surviving from 80 to 100

Compared with the chance of surviving from 80 to 100, it is relatively easy to survive to 80. At least for females in developed countries and in some cases for males as well, median and modal life table life spans are above 80 and for females in some developed countries (such as France and Japan) mean life table life spans (i.e., life expectancy) are now around 83. As shown in Figure 2, only a small fraction of those who reach 80 survive to celebrate their 100th birthday. Some of the data are for the four Nordic countries, so that the trend in survival from 80 to 100 can be examined for a period of more than a century. The astonishing increase in an octogenarian's chances of becoming a centenarian, a more than 10-fold multiplication for both males and females, is due to the substantial fall in mortality between 80 and 100, as documented above. If death rates in some period are R times higher than they were in an earlier period over some age range and if survival over the interval in the more recent period is s, then survival in the earlier period is s raised to the R power. Because of this leverage, moderate changes in death rates can lead to dramatic changes in survival, especially, at ages when survival is low.

Centenarians were so exceedingly rare a couple of hundred years ago that in most countries in most years no one celebrated his or her 100th birthday (Vaupel and Jeune 1995 a, b; Wilmoth 1995). The proliferation of centenarians since then

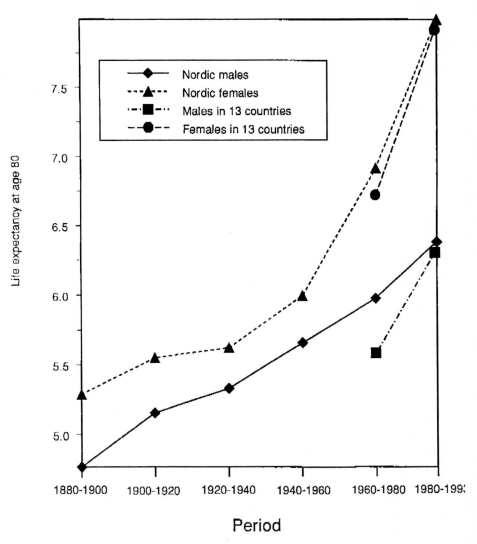

Fig. 3. Life table expectancy at age 80

and especially since the middle of the 20th century is analyzed by Vaupel and Jeune (1995b). As they explain, most of the radical rise in the numbers of centenarians can be attributed to the marked increase in survival from age 80 to 100. The rapid growth of the population of centenarians (by far the most rapidly growing age group in most developed countries, albeit still a small population everywhere) is thus a further reflection of the plasticity of oldest-old mortality.

Figures 3 and 4 depict the rise in remaining life expectancy at ages 80 and 100. Life expectancy is short at age 80 and very short at age 100, but there has been a substantial increase in the time an octogenarian or a centenarian can expect to survive. All things being equal, if remaining life expectancy at some age

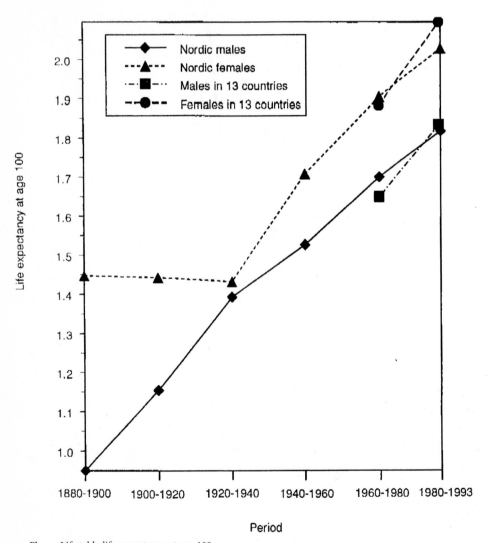

Fig. 4. Life table life expectancy at age 100

increases by some proportion, then population size increases by the same proportion. Hence the increase in life expectancy for 80- and 100 year-old has been a major factor in the growth in the population above ages 80 and 100.

Finally, Figure 5 provides a different perspective on the plasticity of oldest-old mortality by showing the considerable increase in the Nordic countries in the modal life table age at adult death. This measure, which is calculated by finding the age at which the d(x) value in lifetables is maximized, can be considered a measure of the most typical age at death at adult ages or the most common adult life span. In the late 19th century, the modal age hovered around age 75 for males and 76 for females. By the 1940s and 1950s the modal age had risen to about 78

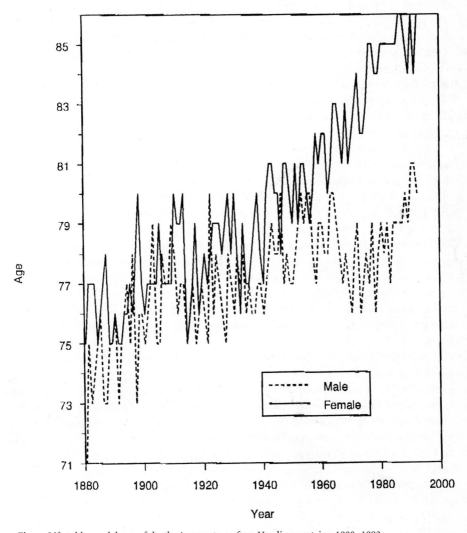

Fig. 5. Life table modal age of death. Aggregate or four Nordic countries, 1880–1993

for males and about 80 for females. Thereafter, the male modal age fell a bit and then rose to a level of 80 or 81 in the 1990s. For females there was a sharp rise in the modal age after 1960, leading to a divergence of female and male values. By the 1990s the female modal age at death hovered between 84 and 86. Over the course of the century from the 1880s and 1890s to the early 1990s, the typical female life span, as calculated from life table values, thus increased by roughly a decade.

Two other measures of the plasticity of oldest-old mortality are worth considering. The first is the age at which remaining life expectancy declines to some level, such as five years. As shown by Vaupel and Lundström (1994) for Swedish females, this age increased by about five years from the 1930s to the 1980s, from 81.5 to 86.7. For males, this increase was about three years, from 80.9 to 83.7. The second measure is the age at which the central death rate reaches some level such as one chance in eight. Vaupel and Lundström's analysis of Swedish data indicates that this age rose from about 81 to about 87 for females and from about 80 to 83 or 84 for males between 1950 and 1990. This age shift in mortality demonstrates that it has been possible to significantly postpone death even among the oldest-old. Information on age shifts in mortality for a variety of countries and at a variety of levels of mortality is presented in Vaupel et al. (1996) and Kannisto (1996).

If it became more and more difficult to make further improvements in mortality as mortality declined, then countries with the lowest levels of mortality might be expected to show the slowest rates of mortality improvement. Furthermore, since female mortality is lower than male mortality, it might be expected that rates of improvement for females would be lower than for males. Data in Kannisto et al. (1994) demonstrate, however, that there is no correlation between the level of oldest-old mortality in developed countries and the rate of improvement. In particular, some countries, such as Japan, that have low levels of mortality also show rapid rates of improvement. Furthermore, in all of the developed countries studied by Kannisto et al. (1994) rates of improvement for females at the oldest-old ages were higher than for males, although levels of mortality for females were lower. This result is consistent with similar findings presented above.

Manton and Vaupel (1995) showed that mortality in the United States after age 80 is lower than in England, France, Sweden, and Japan. The precise extent of the United States advantage is uncertain because of questions about the quality of United States data, but it seems likely that there is an advantage. Manton and Vaupel consider several alternative explanations for this. One possibility is that the United States advantage stems from the quality and extent of health and medical care for the oldest-old in the United States compared with Western Europe or Japan. If so, this would support the hypothesis that oldest-old mortality is amenable to health-care interventions. In any case, there is considerable evidence, briefly reviewed by Manton and Vaupel, that indicates that a variety of interventions can have a major effect in reducing specific causes of death among the very old.

Discussion

Most demographic projections have been and continue to be based on the assumption that mortality is more or less intractable at advanced ages. The results presented above, however, indicate that oldest-old mortality is plastic and has declined substantially over time.

In addition, most projections assume that rates of mortality improvement at older ages decelerate with increasing age. For males after 1960 in the countries analyzed above this was not true: rates of improvement were roughly the same for sexagenarians, septuagenarians, octogenarians, and nonagenarians. For females, however, rates of mortality improvement did tend to decline with age, reaching levels among nonagenarians that were similar to males rates of improvement. If absolute rather than relative mortality improvements are ana-lyzed, then the amount of improvement over time for both male and females is much greater at the oldest ages than among the younger elderly.

Finally, most projections assume that rates of oldest-old mortality improve-ment will decline to zero over time as mortality declines. In fact, for the countries analyzed above, there was a sharp acceleration of rates of improvement.

The questionable assumptions made about oldest-old mortality have been justified by appeal to three related notions:

1) Deaths at older ages are essentially due to old age, and nothing can be done about old age;
2) The typical human organism is not constructed to survive much past age 80 or 90:
3) Causes of death at younger ages are largely extrinsic but causes of death at older ages are mostly intrinsic, and it is very difficult to reduce intrinsic causes of death.

These three notions are dubious, ill-defined speculations, as explained below.

Deaths, even at the most advanced ages, are not due to old age per se; death is due to the diseases, impairments, and injuries that characterize old age. Such health problems often accumulate and multiply with age, so that it may be diffi-cult to ascribe death to a single cause. If progress is made in preventing, postpon-ing, and treating the ill-health conditions of the elderly, oldest-old death rates will decline. As Manton and Vaupel (1995) note, "health changes once accepted as normal features of aging (e.g. frailty and senility) are now viewed as age-related disorders (e.g., osteopososis and the dementias)."

Mortality at advanced ages can only be ascribed to old age per se if there is some biological clock that determines the maximum life span of an individual. Researchers since Buffon (1835) have hypothesized that such a clock exists and that each individual in every species has a specific maximum life span. Evidence from the exciting new field of experimental demography (Carey et al. 1992; Curt-singer et al. 1992) as well as findings from comparative biology (Finch 1990) and from evolutionary theory (Rose 1991) are inconsistent with this hypothesis.

The notion that individuals are not designed to live longer than some span is also inconsistent with the findings of experimental demography and comparative biology. The maximum life span observed in a cohort of some species depends not only on the species but also on the size of the cohort and on environmental conditions. This is true for humans (Vaupel and Jeune 1995; Wilmoth 1995) as well as for other species (Carey et al. 1992; Finch 1990). Evolutionary theory suggests that species are not "designed" to live some amount of time; a species' design depends on opportunities for reproductive success in a very complicated, competitive environment, with post-reproductive survival being of little or no consequence (Charlesworth 1994). Objects that are designed to last to some specific age frequently survive far beyond that age. The Pioneer space probe was designed to reach Mars; it is still functioning far beyond Pluto. Automobiles may be designed to survive their warranty period, but depending on the model and make and how the vehicle is treated, some cars endure for decades.

The concept that some causes of death are intrinsic and others are extrinsic has seemed appealing to many demographers, from Bourgeois-Pichat (1952) to today. It is, however, a concept that is impossible to put into operation unless the most heroic (and questionable) assumptions are made. Accidental deaths and infectious-diseases deaths might seem to be extrinsic, but mortality from accidents and infections rises exponentially with age at older age. Cancer is usually classified as intrinsic, but almost all cancers are due to environmental insults.

The distinction between intrinsic and extrinsic causes of death is made because it is believed that intrinsic causes are intractable. This contention is questionable. Nearsightedness might be classified as an intrinsic disorder because it is largely genetic in origin; normal vision is readily restored by eyeglasses. The risk of Alzheimer's disease depends on the proteins that an individual is genetically programmed to produce as well as on environmental factors such as blows to the head in childhood. It may turn out to be easier to develop pharmaceuticals that induce the body to produce salubrious proteins than to reduce children's propensity to fall.

In sum, the "theory" underlying many demographers' beliefs about oldest-old mortality is questionable, speculative, imprecise, and inconsistent with available evidence. In the absence of strong theory, demographers should do what they have done since Graunt and Halley: rely on careful empirical studies of the rich lodes of population data. The Kannisto-Thatcher Oldest-Old Database and other databases in the Odense Archive of Population Data on Aging, which are freely available for research, now make such demographic analysis feasible. The results presented above and in other publications based on these data suggest that such demographic analysis will require a radical revision of our understanding of mortality at the older ages when death now usually strikes. In particular, the prospects for longer lives no longer seem as remote as many demographers once thought.

This can be illustrated with a simple calculation that returns us full circle to the start of this essay. At current death rates in France for females, the chance of surviving to age 60 is some 92 % and the chance of surviving from 60 to 80 is close to 3 in 4. Enduring another score of years is, however, much harder; only one person in 40 makes it. How then can it be plausible that half of newborn French girls may survive to 100? Consider the following scenario. To be conservative, suppose nothing can be done to reduce the 8 % toll of mortality before age 60. Suppose improvements in mortality among 60- and 70-years-old continue to accumulate at about the current rate of 2 % per year. If so, some 85 % of the 1996 cohort of French girls will still be alive at age 80. At current mortality rates, there is an 85 % chance of surviving to age 70, so this progress would simply add a decade of life to the 85 % who live longest.

Now, however, comes the hazardous segment of life from age 80 to 100. If half of the 1996 cohort is to survive to age 100, then almost three in five will have to make it through their octogenarian and nonaagenarian years, compared with one in 40 at current death rates. This is the stage of life during which the most lives will have to be saved. It turns out, however, that the pace of the required improvement is comparable to the pace achieved in France and other developed countries in recent years. What is necessary at these ages is an average rate of mortality reduction of a bit more than 2 % per year. Alternatively what is needed is for the absolute level of mortality at these ages to decline at the same pace as over the past three decades. This is equivalent to requiring that survival from 80 to 100 continues to improve at the same rate as over the last 30 years. In France female survical from age 80 to 100 has increased by a factor of three since the mid 1960s. If it continues to increase by another factor of three every 30 years for the next 90 years or so, then a bit more than three in five of the 1996 female cohort who reach age 80 will endure to age 100.

It would be injudicious to make too much of these simple calculations. Nonetheless, they do suggest that substantial increases in human longevity are not out of the question. Note that the calculations assume that no further mortality reductions are achieved at ages below age 60. Furthermore, the calculations make no assumption about mortality after age 100. The calculations simply assume that future rates of mortality improvement between ages 60 and 80 and between ages 80 and 100 will average about the same pace as in recent years.

Conclusion

Knowledge of the determinants of human longevity is still sparse (Christensen and Vaupel 1996). On the population level, demographic trends can be analyzed, as above. In addition, a number of factors have been identified that are associated with mortality and with cause-specific mortality, such as cigarette smoking, or that appear to be protective, such as red wine. On the individual level, however, these associations are too weak to be reliable predictors of a person's life span. This can be illustrated by a case history. A Danish woman who was born in 1890 and who grew up in a poor family was sent away from home when she was 15 because she had severe tuberculosis that was a potential hazard to the rest of the family. The infection was treated with an operation. Later she got breast cancer, first in one breast and subsequently in the other, and had both breast removed. The woman herself told this story, which was verified in 1996 shortly after her 105th birthday.

The prospects for deeper understanding of the determinants of longevity may, however, be good. Rapid progress in genetics may add considerably to our understanding of survival. Because the processes of aging in such species as yeast, worms, insects, and rats are similar in some ways to the processes of aging for humans, advances in experimental gerontology may prove to be informative (e.g. Finch 1990; Carey et al. 1992; Curtsinger et al. 1992). Demographic and epidemiological studies of human populations may also play an important role as more reliable and more extensive data are collected and analyzed. Mortality is changing in different regions and countries: life expectancy is declining in Russia and parts of Eastern Europe and increasing in France and most other developed countries. Mortality change in Denmark is following a different pattern than in the other Nordic countries (Härö 1995). In many countries, the changes in mortality rates vary substantially at different ages.

The variety and speed of mortality change provides excellent opportunities for identifying underlying mechanisms and causal factors if appropriate data are gathered, especially data on the specific characteristics of individuals. Particularly in the Nordic countries but also in France, Italy, and elsewhere, the development of large registers of health-related information about individuals – including such special populations as twins, adoptees, and centenarians, as well as the general population – may permit significant development of knowledge about the determinants of the duration of life. Such developments may help inform, as well as being informed by, biomedical research.

If mortality is going to be reduced several fold, continued progress in understanding the determinants of survival and longevity is required. Steady deepening of knowledge, with a cumulative series of research advances, could – and I think probably will – revolutionize how long we live.

References

Ahlburg DA, Land KC (1992) Population forecasting; guest editors introduction. Intnl T Forecasting 8(3):289–299

Ascher W (1978) Forecasting: an appraisal for policy makers and planners. Johns Hopkins University Press, Baltimore

Bourgeois-Pichat J (1952) Essai sur la mortalité ‚biologique' d l'homme. Population 7:381–394

Bourgeois-Pichat J (1978) Future outlook for mortality decline in the world. Pop Bull UN 11:12–41

Buffon GLL (1835) Oeuvres complètes de Buffon. Vol. IV. Paris, P. Duménil, p 108

Carey JR, Liedo P, Orozco D, Vaupel JW (1992) Slowing of mortality rates at older ages in large medfly cohorts. Science 258:457–461

Charlesworth B (1994) Evolution in age-structured populations. New York, Cambridge University Press

Christensen K, Vaupel JW (1996) Determinants of longevity: genetic environmental, and medical factors. J Int Med, in press

Coale A, Guo G (1989) Revised regional model life tables at very low levels of mortality. Pop Index 55:613–643

Curtsinger JW, Fukui HH, Townsend D, Vaupel JW (1992) Demography of Genotypes: Failure of the Limited Lifespan Paradigm in Drosophila melanogaster. Science 258:461–463

Demeny P (1984) A perspective on long-term population growth. Pop Dev Rev 10:103–126

Depoid F (1973) La mortalité des grands viellards. Population 28:755–792

Finch CE (1900) Longevity, senescence, and the genome. Chicago, University of Chicago Press

Härö AS (1995) Surveillance of mortality in the Scandinavian countries 1947–1993. Helsinki, The Social Insurance Institution

Kannisto V (1988) On the survival of centenarians and te span of life. Pop Stud 42:389–406

Kannisto V (1994) Development of oldest-old mortality, 1950–1990: evidence from 28 countries. Odense, Denmark, Odense University Press

Kannisto V (1996) The Advancing Frontier of Survival. Odense, Denmark, Odense University Press

Kannisto V, Lauritsen J, Thatcher AR, Vaupel JW (1994) Reductions in mortality at advanced ages: several decades of evidence from 27 countries. Pop Dev Rev 20:793–810

Keilman N (1995) Forecasting errors. Paper presented at the Workshop of Forecasting, Juensuu, Finland

Keyfitz N (1981) The limits of population forecasting. Population Dev Rev 7(4):579–593

Manton KG, Vaupel JW (1995) Survival after the age of 80 in the United States, Sweden, France, England, and Japan. New Engl J Med 333:1232–1235

Rose M (1991) Evolutionary biology of aging. New York, Oxford University Press.

Stoto M (1983) The accuracy of population projections. J Am Stat Assoc 78:13–20

Thatcher AR (1978) Mortality at the highest ages. J Inst Actuaries 114:327–338

Thatcher AR, Kannisto V, Vaupel JW (1997) The force of mortality from age 80 to 120. Odense, Denmark, Odense University Press

United Nations (1991) United Nations Demographic Yearbook. New York, United Nations

Vaupel JW, Jeune B (eds) (1995a) Exceptional longevity: from prehistory to the present. Odense, Denmark, Odense University Press

Vaupel JW, Jeune B (1995b) The emergence and proliferation of centenarians. In: Vaupel JW, Jeune B (eds) Exceptional longevity: from prehistory to the present. Odense, Denmark, Odense University Press

Vaupel JW, Lundström H (1994) The future of mortality at older ages in developed countries. In: Lutz W (ed) The future population of the world. London, Earthscan Publications.

Vaupel JW, Zhenglian W, Andreev K, Yashin A (1997) Population data at a glance: shaded contour maps of demographic surfaces over age and time. Odense, Denmark, Odense University Press

Vincent P (1951) La Mortalité des Viellards. Population 6:181-204

Wilmoth J (1995) The earliest centenarians: a statistical analysis. In Vaupel JW, Jeune B (eds) Exceptional longevity: from prehistory to the present. Odense, Denmark, Odense University Press

Acknowledgment

The author thanks Kirill Andreev for his assistence in preparing the tables and figures.

Towards a New Horizon in Demographic Trends: The Combined Effects of 150 Years Life Expectancy and New Fertility Models

J. Vallin[*], *G. Caselli*[**]

The hypothesis of a widespread and conclusive stabilization of the world's population, the focus of recent United Nations' forecasts, is obviously merely a reference model – the demographic transition. It is fairly plausible that initially populations will tend to follow this model. However, on the one hand, this is by no means certain and, on the other hand, it is unlikely that after this "transition" things will continue as before. Regarding the first point, the United Nations' forecasts comprise, on both sides of the average hypothesis, "low" and "high" hypotheses whose difference with the former lies in the level to which fertility would tend, with mortality remaining unchanged (United Nations 1995). Among others, the United Nations is studying the consequences of a convergence of fertility levels towards 1.7 children per woman, on the one hand, and towards 2.5 on the other hand, as an alternative to the 2.1 that assures strict generation replacement. In both cases, the assumption that the world population's size will stabilize is no longer valid. A level of 2.5 children per woman corresponds to an infinite growth, whereas with 1.7 children, growth would soon give place to diminution and the population would tend to become extinct.

However, whether with the central hypothesis of a size stabilization or the hypotheses of continuous growth or diminution, the world's population would become a "stable population" stricto sensu, that is a population whose parameters are all constant and whose age structure becomes consequently invariable. In other words, even if the size of the population keeps on changing, it will be, at the end of the "transition", only under the effect of completely frozen behaviours. Of course, this second postulate can also be questioned. It is even very likely that it will be questioned by facts.

Not only is it unlikely that some day human behaviours might definitively freeze, but it seems to us that we can already observe the early beginnings of new changes that might be the signs of the advent of new demographic revolutions. On the one hand, the recent change in mortality at old ages reinforces a number of authors' assumptions that life expectancy could increase far beyond the limit of 85 years adopted in all UN forecasts. On the other hand, the current variations observed in fertility, and particularly the recent evolution of its calendar could

[*] Institut national d'études démographiques, Paris
[**] Dipartimento di Scienze Demografiche Universita degli studi di Roma ,La Sapienza'

J.-M. Robine et al. (Eds.)
Longevity: To the Limits and Beyond
© Springer-Verlag Berlin Heidelberg New York 1997

lead to the adoption of fertility models by age totally different from those which are referred to up to now. In other words, not only is the current transition probably not the only foreseeable demographic evolution, but also new changes might take place even before the current transition that is referred to in the UN forecasts is completely generalized.

To make things clear, we will simplify the problem by ignoring the hypothesis of an overlap between the current transition and the beginning of new changes. We will not set it aside on the grounds of its improbability (given the diversity of current situations and particularly the advance taken by certain populations, we even think that it is more plausible than that of a total separation over time, at the world level), but because it seems to us that it is more useful, from a pedagogical viewpoint, to make a clear distinction between what the results of the current transition (reflected in the UN forecasts) may be, and what an evolution towards fertility and mortality schemes different from those adopted in the framework of these forecasts could bring about.

Therefore we will consider the results of the UN forecasts just as they are, for the year 2050, and we will think out new schemes of fertility and mortality evolution for the subsequent period. Thus, we will adopt a kind of compromise insofar as, in the UN forecasts, the world's population is still far from achieving perfect stability by that time. For example, Africa will be still experiencing strong evolution. However, at the world level, most of the current transition will be achieved by that time. This will prevent us from extending our "fit of futurism" too far.

To begin with, we must think out new possible fertility and mortality evolutions. We will examine, in the next two parts of our work, the consequences they would entail in the far future of the world's population, in terms of global size as well as of age structure.

Thinking out New Fertility and Mortality Evolutions

As early as 1966, during the 1st European Demographic Conference convened by the Council of Europe even before the prospect of a generalization of the demographic transition became plausible enough to be adopted as the central hypothesis of the UN forecasts, Jean Bourgeois-Pichat assumed that progress in the field of biology would some time or other bring about new schemes of demographic behaviours. Not only would increased longevity – becoming a factor of the population's aging – compel future societies to reconsider the distribution of the economic and social roles played by the different age cohorts, but also delaying menopause would make it possible for women to spread the constitution of their descent over time. Thus he advanced the idea that the day might come when couples could procreate in two stages, engendering two children about age 28 and, once again, about age 50 after bringing up their first brood. He asserted that this would raise a great problem of too rapid population growth for societies, which would have to decide on who could have access to this second stage of fertility life (Bourgeois-Pichat 1966). Some twenty years later, Jean Bourgeois-Pichat took up this idea again, relaying the very optimistic convictions of some biologists, such as Roy Walford (1984), about the future of life expectancy, and adapting his reflections on fertility to the new context brought about by the fertility decline in Europe. Overpopulation is no longer to be feared but it is necessary to find a means to reach strict generation replacement again thanks to a two-stage fertility (Bourgeois-Pichat 1987). Of course, this scenario is not the only possible one. As regards the very-long-term future of European population, Jean Bourgeois-Pichat thought out two possible alternatives to the UN forecast: the first one shifting life expectancy to 100 years, with a fertility level maintained at 2.1 children per women, the other foreseeing a continuous fertility diminution, up to the population's extinction (Bourgeois-Pichat 1988).

At a continent's scale, and still more at a country's scale, such scenarios, to be complete, should also comprise hypotheses on migrations. At the level of the world's population, on which we are focusing in our study, it is of course unnecessary, except if we intend to also make an account of the evolution of geopolitical relations. We will limit our study, as Bourgeois-Pichat did, to speculating about the specific effects of fertility and mortality variations.

Towards New Fertility Models

In addition to the three basic hypotheses used by the UN, according to which all populations are converging towards a fertility of 1.7, 2.1 or 2.5 children per woman, we will think out other fertility models, completely different, that are liable to be adopted.

The Transition to the Only Child

The first model we will consider is that of the only child. It is this model that the Chinese government tried to enforce in China and, even if this policy has not been as successful as expected, it is interesting to determine the consequences its achievement would bring about. This is all the more true as it seems that this model could become the rule, at least temporarily, in some western European countries. For example, the total fertility rate of some Italian regions has recently fallen to less than 1 (0.7 in Emilia, 0.8 in Liguria; Santini 1995; Zanatta and De Rose 1995). Several fertility models with one child per woman are of course possible. In this study, we use a rather precocious model, the calendar of which is copied from the one adopted by the UN for its fertility hypothesis of 1.7 children per woman. Figure 1 illustrates fertility rates by quinquennial age group corresponding to the three UN hypotheses (2.1, 1.7 and 2.5) and to the only-child hypothesis.

The transition to the only child as described above, which is based on the UN 1.7-children model, takes up to 35 years. Intermediary stages are obtained through linear interpolations (Fig. 2).

Two-Mode Fertility

The second fertility hypothesis we will keep here is directly inspired from Jean Bourgeois-Pichat's idea of a two-stage procreation. However, it does not mean a return to high fertility, as the author suggested when he brought forward this hypothesis for the first time. On the contrary, it is a means to keep up strict generation replacement in the future. Of course, such a hypothesis is actually valid only in the case of an important increase in life expectancy. It does not only suggest that menopause would be much postponed but also that women would live in good health long enough to bring up their second brood. Therefore, the general philosophy of this type of hypothesis is that of a new transition towards a new equilibrium.

In this case also, there is an infinity of possible bimodal models and as many ways to change from the "current" 2.1 children per woman hypothesis (UN average hypothesis) to the new model selected. In this case, we suppose that women assure descent by combining a first fertile period during which they give birth to 1.4 children on average and a second period, 30 years later, during which they give birth to 0,7 child on average, each time using a calendar copied from the age distribution characterizing the UN average hypothesis. However, we will keep two different schemes, shifting forward the beginning of fertile life. In the first

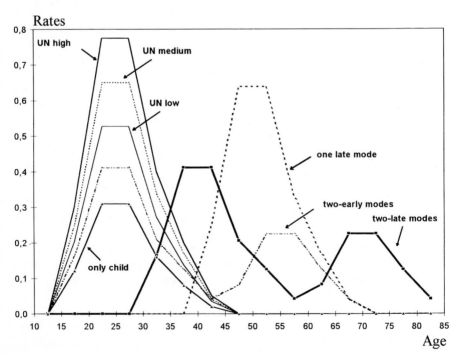

Fig. 1. Fertility rates, by quinquennial group, for the three UN hypotheses (2.5, 2.1 and 1.7 children per woman) and for the four new hypotheses explored here.

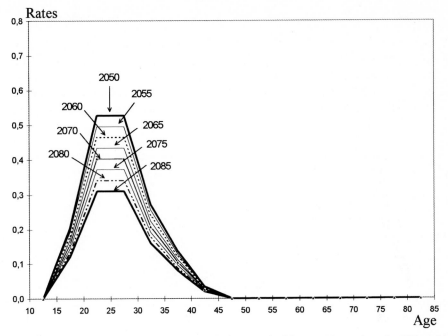

Fig. 2. Change in fertility rates by age group in the framework of the transition to the only-child model within 35 years, set up from the UN low hypothesis (1.7 children per woman).

scheme, called they "two-early-mode model," fertile life begins at age 15, like in the UN hypothesis, then goes through a first mode between ages 20 and 29. The second mode takes place from age 50 to age 59. In the second case, called the "two-late-mode model," the whole calendar is posponed by 15 years, with a first model at ages 35–44 and the second at ages 65–74 (Fig. 3).

As regards the transition from the starting model to the new one, we kept the principle of a progressive adoption of the new model by younger generations. Thus, fertility at young ages begins to diminish before the second period of fertile life at older ages starts. Thus, the total fertility rate indicator first diminishes strongly before going back to its balance level (set here at 2.06 to settle on a quasi suppression of mortality before the entry into fertile life, at the end of a 50-year evolution) (Tables 1 and 2).

Of course this diminution is more important in the case of the transition to two-late modes: the total fertility rate falls then to 0.86 after 15 years, instead of 1.45 in the case of the transition to two-early modes.

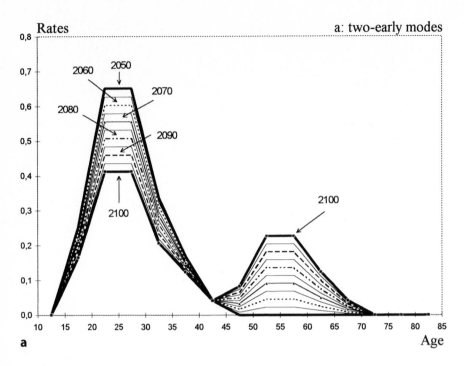

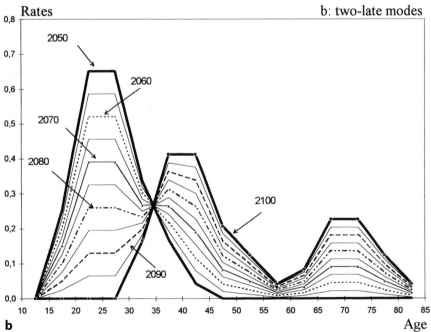

Fig. 3. Evolution of fertility rates by age group in the framework of the transition from the UN average hypothesis (2.1 children per woman) to a two-mode model (a: early; b: late).

Table 1. Evolution of fertility rates by quinquennial age group, in the framework of the transition from the UN average hypothesis to the two-early-mode model

Age	Year										
	2050	2055	2060	2065	2070	2075	2080	2085	2090	2095	2100
15–19	0.252	0.223	0.194	0.165	0.165	0.165	0.165	0.165	0.165	0.165	0.165
20–24	0.651	0.571	0.492	0.412	0.412	0.412	0.412	0.412	0.412	0.412	0.412
25–29	0.651	0.571	0.492	0.412	0.412	0.412	0.412	0.412	0.412	0.412	0.412
30–34	0.336	0.293	0.249	0.206	0.206	0.206	0.206	0.206	0.206	0.206	0.206
35–39	0.168	0.153	0.138	0.124	0.124	0.124	0.124	0.124	0.124	0.124	0.124
40–44	0.042	0.042	0.042	0.042	0.042	0.042	0.042	0.041	0.041	0.041	0.041
45–49	0.000	0.000	0.000	0.010	0.022	0.034	0.046	0.057	0.069	0.081	0.082
50–54	0.000	0.000	0.000	0.028	0.061	0.093	0.125	0.158	0.190	0.223	0.227
55–59	0.000	0.000	0.000	0.028	0.061	0.093	0.125	0.158	0.190	0.223	0.227
60–64	0.000	0.000	0.000	0.015	0.033	0.051	0.068	0.086	0.104	0.121	0.124
65–69	0.000	0.000	0.000	0.005	0.011	0.017	0.023	0.029	0.035	0.040	0.041
CFI*	2.1	1.85	1.61	1.45	1.55	1.65	1.75	1.85	1.95	2.05	2.06

* Conjonctural Fertility Index

Late Fertility

We will formulate a last fertility hypothesis by pushing to the extreme the assumption that the diminution of the cyclical indicator observed during the last decades in western Europe is the consequence of the transition from early fertile life to a later calendar. From the 1.7 children of the UN low hypothesis, we will find 50 years later, the 2.06 children that will assure the equilibrium (given the

Table 2. Evolution of fertility rates by quinquennial age group in the framework of the transition from the UN average hypothesis to the two-late-mode model.

Age	Year										
	2050	2055	2060	2065	2070	2075	2080	2085	2090	2095	2100
15–19	0.252	0.168	0.084	0.000	0.000	0.000	0.000	0.000	0.000	0.000	0.000
20–24	0.651	0.434	0.217	0.000	0.000	0.000	0.000	0.000	0.000	0.000	0.000
25–29	0.651	0.434	0.217	0.000	0.000	0.000	0.000	0.000	0.000	0.000	0.000
30–34	0.336	0.279	0.222	0.165	0.165	0.165	0.165	0.165	0.165	0.165	0.165
35–39	0.168	0.249	0.331	0.412	0.412	0.412	0.412	0.412	0.412	0.412	0.412
40–44	0.042	0.079	0.116	0.153	0.190	0.227	0.264	0.301	0.338	0.375	0.412
45–49	0.000	0.000	0.000	0.026	0.055	0.085	0.114	0.143	0.173	0.202	0.206
50–54	0.000	0.000	0.000	0.015	0.033	0.051	0.068	0.086	0.104	0.121	0.124
55–59	0.000	0.000	0.000	0.005	0.011	0.017	0.023	0.029	0.035	0.040	0.041
60–64	0.000	0.000	0.000	0.010	0.022	0.034	0.046	0.057	0.069	0.081	0.082
65–69	0.000	0.000	0.000	0.028	0.061	0.093	0.125	0.158	0.190	0.223	0.227
70–74	0.000	0.000	0.000	0.028	0.061	0.093	0.125	0.158	0.190	0.223	0.227
75–79	0.000	0.000	0.000	0.15	0.033	0.051	0.068	0.086	0.104	0.121	0.124
80–84	0.000	0.000	0.000	0.005	0.011	0.017	0.023	0.029	0.035	0.040	0.041
CFI*	2.10	1.64	1.19	0.86	1.05	1.24	1.43	1.62	1.81	2.00	2.06

* Conjonctural Fertility Index

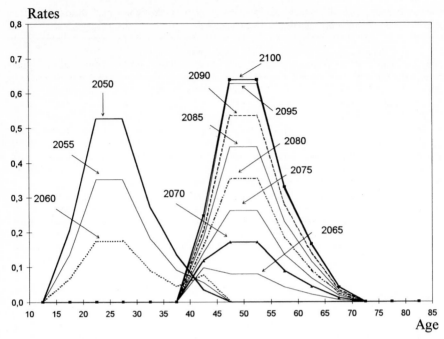

Fig. 4. Evolution of fertility rates by age group of the UN low hypothesis (1.7 children per woman) in the transition to a one-late-mode model restoring the level of equilibrium (2.06 children per woman).

suppression of mortality at younger ages), but with a much later fertile life reaching its higher point at age 45–54 instead of 20–29 (Fig. 4).

In the meantime, of course, the cyclical indicator will have gone through a minimum value, still lower than in the two-late-mode model: 0.32 child per woman (Table 3)! In fact, we could have carried the paradox to make the current fertility equal to zero for a few years.

Figure 5 sums up the different evolutions under consideration as regards fertility.

What Limits to Life Expectancy Increase:

The prospects of life expectancy progress are now a very much debated question. Everybody agrees that average life duration should still increase during the next decades, considering the recent evolutions resolutely towards an increase in the most advanced countries. But up to which limits? Here lies the disagreement.

From the early 1950s, Jean Bourgeois-Pichat calculated a "limit of biological mortality table" showing the life expectancy that could be reached if all "exogenous" causes of death could be eliminated. These causes, due to clearly identified pathogenic agents or to external causes, could be avoided thanks to the medical

Table 3. Evolution of fertility rates by quinquennial age group in the framework of the transition from the UN low hypothesis to the "one-late-mode model."

Age	Year										
	2050	2055	2060	2065	2070	2075	2080	2085	2090	2095	2100
15–19	0.204	0.136	0.068	0.000	0.000	0.000	0.000	0.000	0.000	0.000	0.000
20–24	0.527	0.351	0.176	0.000	0.000	0.000	0.000	0.000	0.000	0.000	0.000
25–29	0.527	0.351	0.176	0.000	0.000	0.000	0.000	0.000	0.000	0.000	0.000
30–34	0.272	0.181	0.091	0.000	0.000	0.000	0.000	0.000	0.000	0.000	0.000
35–39	0.136	0.091	0.045	0.000	0.000	0.000	0.000	0.000	0.000	0.000	0.000
40–44	0.034	0.055	0.077	0.098	0.119	0.141	0.162	0.183	0.205	0.226	0.247
45–49	0.000	0.000	0.000	0.080	0.171	0.262	0.354	0.445	0.536	0.627	0.639
50–54	0.000	0.000	0.000	0.080	0.171	0.262	0.354	0.445	0.536	0.627	0.639
55–59	0.000	0.000	0.000	0.041	0.088	0.135	0.182	0.230	0.277	0.324	0.330
60–64	0.000	0.000	0.000	0.021	0.044	0.068	0.091	0.115	0.138	0.162	0.165
65–69	0.000	0.000	0.000	0.05	0.011	0.017	0.023	0.029	0.035	0.040	0.041
CFI*	1.70	1.17	0.63	0.32	0.60	0.89	1.17	1.45	1.73	2.01	2.06

* Conjonctural Fertility Index

advancements realized, provided that the whole population could have access to them. Thus he obtained a maximum life expectancy of 76 years for men and 78 years for women (Bourgeois-Pichat 1952). For a long time, the UN adopted a level close to this maximum (75 years) as the level towards which all populations

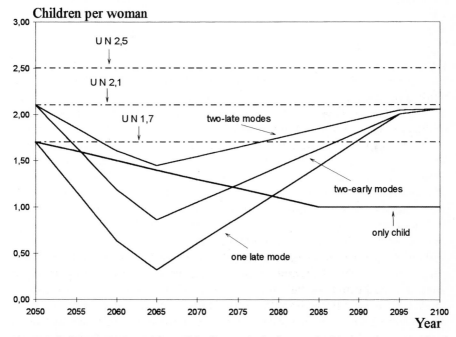

Fig. 5. Evolution over 50 years of the total fertility rate in the framework of the hypotheses considered here, compared to the UN hypotheses.

were supposed to tend, in its forecasts of the world's population (United Nations 1982).

Nowadays, it is clear that reality has exceeded fiction, at least as regards women, as women's life expectancy is already above 80 years in some 10 industrial countries (Australia, Canada, Spain, France, Iceland, Italy, Norway, the Netherlands, Sweden, Switzerland) and is even above 82 years in Japan (Guibert-Lantoine and Monnier 1995). The question is: Is this due to the fact that pathogenic agents, up to now unidentified or badly identified, and the means, to fight against them, have been discovered, or is recent progress in life expectancy due to a recession of "endogenous" mortality itself, which, according to Jean Bourgeois-Pichat, is due to the biological constitution of the human organism and on which he thought it would be far more difficult to act?

In the first case, the possible limit would not have changed but it would be just a little higher than Jean Bourgeois-Pichat asserted in 1952. In this case, we would just have to take up the calculations and adapt them to our present level of knowledge about causes of death. Jean Bourgeois-Pichat published, in 1978, an updating of his limit of biological life table. However, he contented himself with using more recent statistical data, without changing the distribution between exogenous and endogenous causes that he had followed in 1952, which explains the paradoxical characteristic of his new results, which show an upper limit clearly more important for women (80.5 years instead of 78.2) but inferior for men (73.8 years instead of 76.3). By maintaining all cancers in the endogenous category, he ascribed to a biological cause the invading effect of an external cause, addiction to smoking (Bourgeois-Pichat 1978; Vallin 1994). A few years later, Bernard Benjamin (1982), taking up the distinction between endogenous and exogenous but taking into account the most recent knowledge about morbidity factors, obtained a significantly superior limit of life expectancy, about 85 years for both sexes. This result is very similar to that which can be obtained if "diseases consequent on degeneration" (Vallin and Meslé 1988) only are adopted as causes of death. More recently, Olshansky et al. (1990), notably using the works of biologist James Fries (1989), considered as implausible the idea that life expectancy might go beyond age 85; it will go on increasing for some time following the classical hypothesis of a rectangularization of the survival curve, but will necessarily bump against a maximum which, given the unchangeable longevity of human species, can hardly go beyond 85 years. It is this maximum limit of 85 years that the UN have adopted for its forecasts since the early 1980s.

However, it is possible to draw up another interpretation of the recent evolutions. Haven't we indeed lately found the means to fight efficiently against Man's biological aging process and to increase his longevity? In other words, in a field where up to now we could only enable an increasing number of individuals to approach the possible biological limit, aren't we now in the process of shifting this limit? This is what Roy Walford (1982) has been assuming for a long time. This hypothesis has not yet proved true, but it draws more and more arguments from the factual evolutions. On the one hand, the duly stated extreme age of life has lately significantly increased. Jeanne Calment is probably the first woman in the world who ever overpassed age 120 (Allard et al. 1994). But, on the other hand – and this is undoubtedly most meaningful – mortality rates at very old ages are for the first time significantly decreasing (Kannisto 1994). In other words, we may be at the very beginning of a true revolution in the field of human longevity. Taking this revolution as a basis, Walford asserted that life expectancy at about 150 years in the course of the next century is perfectly plausible (Walford 1984).

However, estimating the maximum life expectancy at 150 years is not enough to let us know the new horizon of human life. We still have to discover what the survival curve is leading to this limit, which is still now beyond all understanding. To begin with, if the biological limit is gone over, why should we set another one? And on what basis? Therefore we use this figure just as a point of reference without judging what its nature might be, only to try to determine what its consequences would be. We have then to choose between two possibilities. Either 150-year life expectancy is reached thanks to a technological leap that could immediately benefit the greatest number – in other words, without leaving the scheme of the rectangularization of the survival curve in which we have been for two centuries, or we reach the same result thanks to a speeding up of the mortality decrease at very old ages, bringing about a rapid extension of the age at death and a derectangularization of the survival curve.

Maintaining the rectangularization of the survival curve...?

As an illustration of the first hypothesis, we simply decided to progressively shift the survival curve corresponding to the UN limit table towards increasingly old ages. Thus, the survival at 15 of the UN table becomes 10 years later, the survival at 20, ten years later, the survival at 25, and so on. However, we had to deal with the proportions of survivors before age 15 in a different manner; we had them progressively increase, up to no mortality at young ages. Finally, we completed the UN table, whose limit age is 100, extrapolating quinquennial quotients so as to reach 1 at age 135, before shifting the corresponding proportions of survivors. Figure 6 illustrates the progression of the female survival curve thus obtained.

This procedure leads us from the UN 85-year life expectancy to a 150-year life expectancy for both sexes, over 140 years. In other words, if we start from the result obtained by the UN for 2050, we will have to wait until 2190 to reach the new limit made possible by the technological leap considered. This is assuredly a revolution, although it takes a long time. We prefer this cautious hypothesis to a faster evolution. In fact, population forecasts require that we work by sex. This leads us to make men's life expectancy shift from 82.5 years (UN limit) to 147.4 years and that of women from 87.5 to 152.5 years (Table 4).

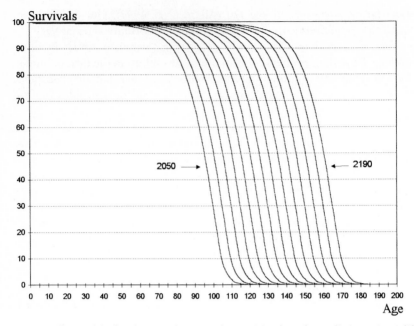

Fig. 6. Evolution of the female survival curve in the transition from the UN limit/maximum life expectancy (87.5 years) to 153-year life expectancy, with preservation of the "rectangularization".

... or extension of the age at death?

The alternative hypothesis of a derectangularization of the survival curve is rendered in a manner just as simple: all quinquennial mortality quotients are reduced by 10 % every five years, and consequently the age at which the quotient reaches 1 is also reduced by five years. Of course, this proportional lowering is more profitable to the ages at which mortality is the highest. This is largely suffi-

Table 4. Evolution of life expectancy in the hypothesis of a transition to 150 years, according to the type of transformation of the survival curve.

Sexe	Year														
	2050	2060	2070	2080	2090	2100	2110	2120	2130	2140	2150	2160	2170	2180	2190
With rectangularity of the survival curve maintained															
Male	82.5	87.5	92.5	92.5	97.5	102.5	107.4	112.4	117.4	122.4	127.4	132.4	137.4	142.4	147.4
Female	87.6	92.6	97.6	97.6	102.6	107.5	112.5	117.5	122.5	127.5	132.5	137.5	142.5	147.5	152.5
Both	85.1	90.0	95.0	95.0	100.0	105.0	110.0	115.0	120.0	125.0	130.0	135.0	140.0	145.0	150.0
With extension of the age at death															
Male	82.5	85.1	87.8	90.7	94.0	97.7	101.9	106.6	111.9	117.8	124.2	131.2	138.6	146.4	150.5
Female	87.6	90.0	92.5	95.3	98.5	102.1	106.1	110.7	115.8	121.4	127.6	134.3	141.5	149.1	153.0
Both	85.1	87.5	90.1	93.0	96.3	99.9	104.0	108.6	113.8	119.6	125.9	132.7	140.0	147.7	151.7

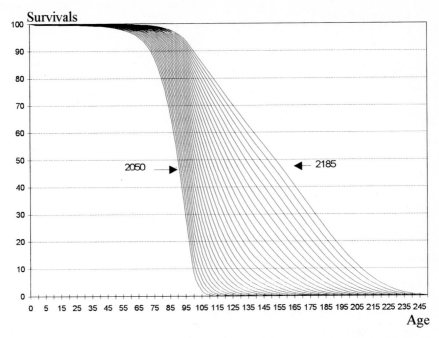

Fig. 7. Evolution of women's survival curve in the framework of the transition from the UN maximum life expectancy (87.5 years) to a 153-year life expectancy, with an "extension of the age at death".

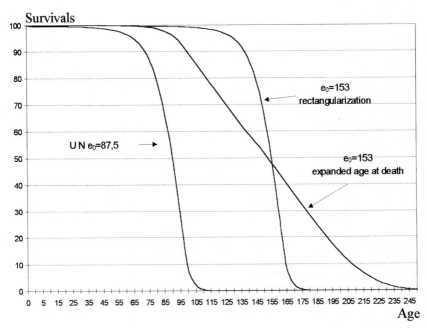

Fig. 8. Comparison of two female survival curves, corresponding to a 153-year life expectancy, to the UN 87.5 years limit.

Life expectancy

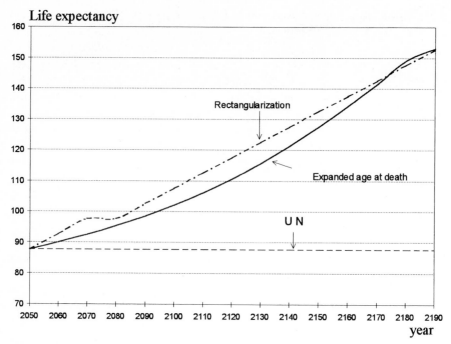

Fig. 9. Evolution of women's life expectancy in the two hypotheses of a transition to 153 years, compared to the UN upper limit.

cient to substantially derectangularize the survival curve. Thus we reach 150-year life expectancy within the same period of time as in the previous scenario, but with a very different extreme age of life: the last deaths occur at age 270 and age 275 instead of 190 and 195 in the preceding case. Figure 7 illustrates this process for women.

Figure 8 compares, also for women, the three mortality limit tables used for the population forecasts mentioned below: the UN limit table and our new tables with 150-year life expectancy.

Finally, Figure 9 illustrates the life expectancy evolution described above. Although they tend towards the same final result, our two extrapolations to 150 years (or more exactly to 153 as they are focused on women) follow rather different courses. The extension of the age at death in a first stage induces a slower progression of life expectancy, but this accelerates in a second stage, and in fine comes up to the level reached in the other hypothesis.

Some possible far futures with life expectancy limited to 85 years

In addition to the three UN fertility hypotheses and the only mortality hypothesis, we have now four new fertility hypotheses and two new mortality hypotheses, that is 21 possible scenarios to explore the long-term future of the world's population. We will first study the consequences of different fertility hypotheses in the framework of a life expectancy with an upper limit fixed at 85 years. The results for the UN average, high and low hypotheses until 2150 are already known, but it is interesting to prolong them beyond this limit in order to compare them to those that would be induced by the achievement of the new hypotheses.

We explored the future until 2315. This may seem to be extravagant, but we will see that, in certain cases, it is still very insufficient to reach stabilization. It is of course with the UN average hypothesis that stabilization will take place most quickly, and in this case the final result will not be far from what was already forecast by the UN for the year 2150. With the high hypothesis, it is an infinite growth that is coming on, but according to what tempo? The achievement of the low hypothesis, and still more the transition to the only child, would bring about the population's extinction. We will explore first these divergent hypotheses. Second, we will examine the more complex results of the upheaval that would be induced by the adoption of radically different fertility models.

Infinite growth or population extinction?

Figure 10 illustrates the evolution of the world's population's total size between 2050 and 2315, with life expectancy fixed at 85 years and four fertility hypotheses: maintenance of 2.5, 2.1 or 1.7 children per woman, already reached in the UN forecasts, or transition to the only child. In the first three cases, connecting with the UN forecasts explains the slight initial difference (respectively, 12, 10 and 8 billion world inhabitants in 2050). In the case of the transition to the only child, we start from the 8 billion of the UN low hypothesis, as it is from this date that we make the average number of children per woman shift from 1.7 to 1.0.

These results are organized around those provided by the perspective of infinite maintenance of the UN model of fertility at 2.1 children per woman. In fact, this last hypothesis does not assure the strict maintenance of the world population's size at the 10 billion reached by 2050, for two reasons. On the one hand, at this date, stabilization is not yet completely achieved and the UN projection forecasts the continuance of some growth over several decades, leading to 12.1 billion by 2150. But, on the other hand, 2.1 children per woman is slightly more than is necessary to reach a perfectly stationary state. Thus we reach 13.5 billion by 2315.

By the same date, maintaining fertility at 2.5, a level slightly above the generation replacement level, leads to 79 billion. This is large but it is by no means extravagant: we can remark that simply maintaining the current fertility rates of the different regions of the world until the UN forecasts' term would lead to 700

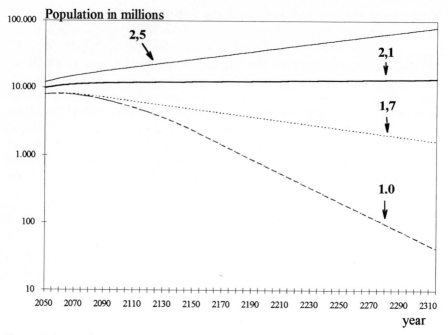

Fig. 10. Evolution of the world's population from 2050 to 2315, with 85-year life expectancy, depending on whether fertility stabilizes at 2.5, 2.1, 1.7 or 1.0 children per woman.

billions as early as 2150 (United Nations 1992; Vallin 1994)! This means that, after the paroxystic stage of the demographic growth characterizing the current transition, prolonged maintenance of fertility at this level, which is slightly higher than the replacement level, would remain far from the current growth rhythm. However, we can note that such maintenance means unlimited growth: prolonging the perspective over time would lead to absurd figures for the population size. But how could we judge now that reaching a population size, however huge it may be, in several hundreds, indeed thousands of years, is absurd? By this time, would not man have had the time to change a lot of things?

It is undoubtedly more interesting to study what maintaining fertility at lower levels than the replacement level would induce. With 1.7 children per woman, we already know that the world population's size would fall to 4 billion in 2150, but a century and a half later, it would be only 1.6 billion, that is 50 times less than with 2.5 children per woman! In this case also, the result is indeed probably less extravagant than we could have imagined. Aren't some environmentalists praying for a return to such size, which is nothing other than the size of the early 20th century? Maintaining sine die fertility below 2.1 necessarily leads to the extinction of humanhood, but with 1.7 children per woman, we can wait and see!

However the transition to the only child would hasten the process. The world population's size would fall below 1 billion around 2185, and by 2315 it would be reduced to 42 millions. This would resemble a return to the Neolithic Age. However, we should note that, even in these conditions, extinction is not foreseeable in the immediate future! The people who strive to find, in this kind of threat, an argument in favour of populationist policies are obviously wrong. The threat would suppose an incredible perseverance from the future generations' part in preferring the only child.

But the most concrete results of such hypotheses lie perhaps less in the evolution of the global size than in that of the population's age structure. Figure 11 illustrates the different profiles that could be obtained by the year 2315. With 2.5 children per woman, the population keeps a rather young profile (29 % of people aged less than 20), which the current European populations forgot already some time ago. Maintaining fertility at the replacement level obviously leads to a profile directly deducted from the survival curve of the mortality table, with 24 % of people aged under age 20, 46 % at ages 20–60, and 30 % above age 60.

A level of 1.7 children per woman and, worse, the transition to the only child, results in a dramatic aging of the age pyramid. In the 1.7 children hypothesis,

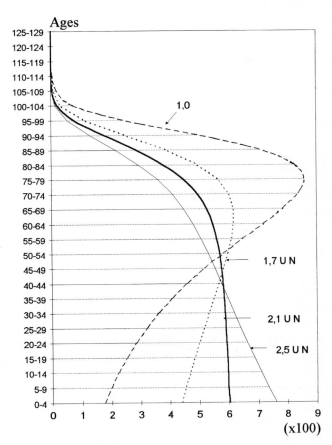

Fig. 11. Population profiles by age reached by 2315 with 85-year life expectancy, depending on whether the fertility stabilizes at 2.5, 2.1, 1.7 or 1.0 children per woman.

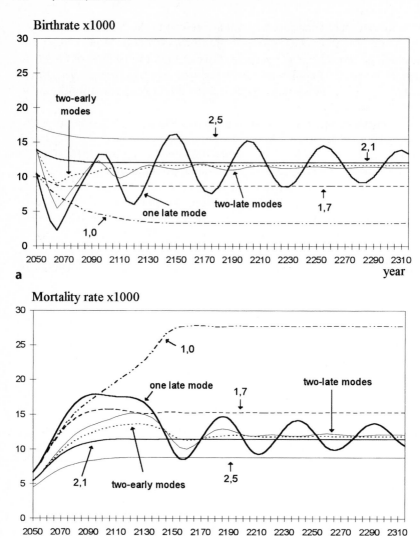

Fig. 12. Evolution of the parameters of the natural increase corresponding to the seven fertility hypotheses (with 85-year life expectancy): **a**) crude birth rate, **b**) crude death rate, **c**) rate of natural increase.

35 % of the 1.6 billion men and woman living in 2315 would be above age 60 and only 22 % would be below age 20. With the transition to the only child, 24 (56 %) of the 42 million people still in existence would be above age 60 and 10 (24 %) would be above age 80. In this case, we are diametrically opposite to the Neolithic Age, in which people over age 80 could be counted on the fingers of one hand.

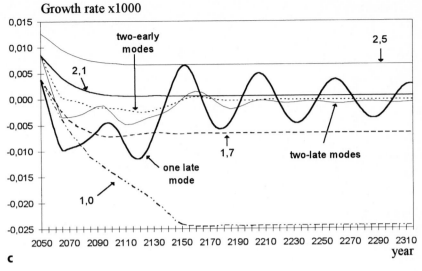

Growth rate x1000

c

Fig. 12c

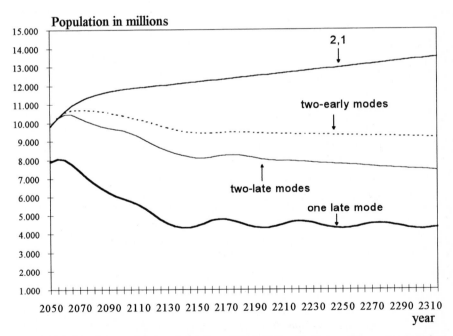

Population in millions

Fig. 13. Evolution of the world's population from 2050 to 2315, with 85-year life expectancy, and 2.1 children per woman, according to the fertility model adopted.

48   J. Vallin, G. Caselli

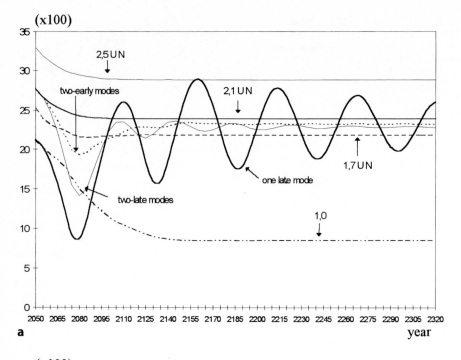

a

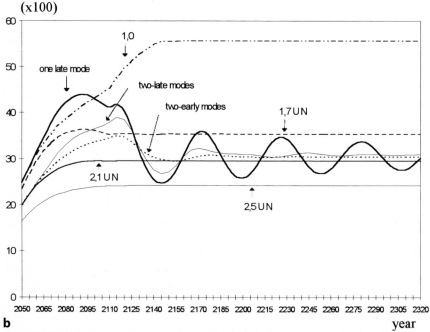

b

Fig. 14. Evolution of the age structure of the world's population, from 2050 to 2315, with 85-year life expectancy, and according to the seven fertility hypotheses: **a)** proportion of people below age 20, **b)** proportion of people above age 60.

Fig. 15. Population profiles by age reached by 2315 with 85-year life expectancy and 2.1 children per woman, according to the fertility model adopted.

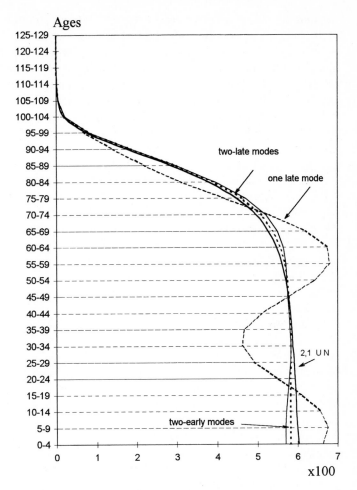

Consequences of the adoption of new fertility schemes

Hitherto, nothing original has been presented compared to the UN forecasts; we just pushed their consequences a little farther. The hypothesis of the transition to the only child itself, as well as its consequences, is rather conventional. But things are quite different if we change the calendar of births in a significant proportion, either by introducing the notion of a two-mode calendar or just by devising a new one-late-mode model. To understand better, we deliberately chose to remain then in the framework of a long-term fertility stabilization at generation replacement level.

Contrary to the preceding situations, in which the parameters of the natural increase (crude birth and mortality rates) rapidly stabilized, stabilization occurs far later in the three following cases, due to the great changes brought about (Fig. 12a, b, c). In the three UN hypotheses, the rate of natural increase was reaching its definitive level (respectively 0.07, 0.7 and –0.7 % per year) as early as 2130. The transition to the only child required some 20 % more years, at the end of which it fell drastically to –2.4 %. In the hypothesis of the transition to bi-modal fertility, the parameter stabilization takes place far later. If the two modes are early, it takes place around 2250, with some hundred years' delay; if they are late, stabilization occurs outside the framework of our forecasts. Finally, if we sig-nificantly postpone all the fertility by setting up a unique mode around age 50, the upheaval induced is so great that birth and mortality fluctuations are still very important at the end of the forecast.

On the contrary, as all these hypotheses converge towards 2.1 children per woman, thus assuring generation replacement at term, the total population's size is far less different (Fig. 13). By 2315, instead of the 13.5 billion resulting from the UN average hypothesis, we would reach 9.2 billion with an early bimodal fertility, 7.4 billion with two-late-modes, and 4.3 billion with one-late-mode. The differ-ences are due to a more or less important diminution stage corresponding to the time required for the adoption of the new fertility model. In this case, the impact of the fluctuations inflicted to birth and mortality rates, particularly discernible in the case of the transition to a late mode, can also be noted in the different tra-jectories.

The contrast between the different population prospects is still more striking when we compare the trajectories by age group (Fig. 14 a, b). In reality, the extreme divergence between the proportions of people aged below 20 and above 60 already mentioned above in the case of the hypotheses of 2.5, 1.7 and 1.0 chil-dren per woman, reaches its maximum very quickly. The proportions are stabil-ized as early as 2110 for the UN high and low hypotheses and, somewhat later, around 2140, in the case of the transition to the only child. On the other hand, the three last hypotheses obviously converge towards the same result as the UN average hypothesis, but through a long period of fluctuations, particularly wide in the case of the transition to a one-late-mode fertility.

At the beginning of the 24th century, that is at the end of our forecast, the profiles per age brought about by both two-mode fertility hypotheses are not yet completely merging with those achieved for long profile of the UN average hypo-thesis (Fig. 15). The profile resulting from the transition to a one-late-mode remains very far from the latter, bearing the trace – almost quite new – of a jolt blown two centuries earlier.

Consequences of Technological Innovations Leading to a 150-year Life Expectancy

In the face of these possible changes in fertility, we can wonder what a lengthen-ing of life expectancy to 150 years would bring about. Would the shape of the new survival curve influence the population evolution?

Evolution of the Total Population

During the whole stage of life expectancy increase, of course, an additional population growth, at equal fertility, takes place. However, the rhythms and width of this phenomenon vary according to circumstances. At the end of our forecast (2315), the total population obtained is grosso modo twice as large as

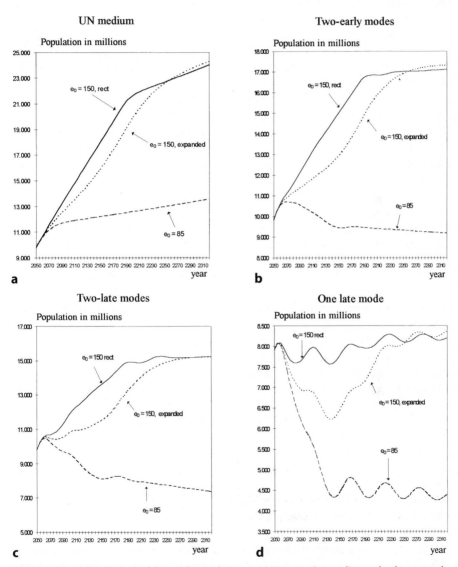

Fig. 16. Compared evolutions of th world's population size (2050–2315) according to the three mortality hypotheses, in the four cases in which fertility stabilizes at 2.1 children per woman: **A)** UN average hypothesis, **B)** two early modes, **C)** two late modes, **D)** one late mode.

that obtained with an 85-year expectancy in the four hypotheses in which fertility becomes stable at 2.1 children per woman: some 24 billion instead of 13 in the UN average hypothesis; 17, 15 and 8 billion instead of 9, 7 and 4 if we adopt a new fertility calendar (respectively, two-early modes, two-late-modes and one-late-mode). In other words, in the prospect of a stabilization of the population size, the transition from 85- to 150-year life expectancy doubles the level at which this stabilization takes place. The result does not vary much whether rectangularity of the survival curve is maintained or not (Addendum 1).

In the UN high hypothesis – the only one forecasting a boundless population growth – the relative effect, of the transition to a 150-year life expectancy is smaller. In 2315, instead of 79 billion inhabitants, we should reach some 118 billion with, this time again, a negligible difference according to the shape of the survival curve adopted.

In the two last fertility hypotheses, leading to decrease, the transition to a 150-year life expectancy changes the result obtained by 2315 in a more significant way. With 1.7 children per woman, we reach a little more than the double of the first result (3.7 to 4 billion instead of 1.6), with also a rather significant change of about 10 % in favour of the scenario of the extension of the age at death. In the case of the transition to the only child, this twofold difference is still more radical: the final result is multiplied by 5 (230 million instead of 42) if the survival curve remains rectangular and by nearly 10 (374 million) if there is an extension of the age at death!

These differences in the results at term go hand in hand with trajectory differences. In the case of a fertility stabilization at 2.5 children per woman, these trajectories are not surprising. Whatever the mortality hypothesis may be, the population size regularly rises and with the exponential rhythm prevailing over the rest, both curves corresponding to 150-year life expectancy are almost superimposed.

On the contrary, in the four fertility hypotheses assuring generation replacement at term, the population doubling, due to the transition to a 150-year life expectancy, is obtained much quicker when the rectangularity of the survival curve is maintained than with an extension of the age at death. The four graphs of Figure 16 clearly illustrate this phenomenon by putting to the fore a period of one hundred years during which both trajectories are significantly distant. The case of the transition to one-late-mode fertility is particularly interesting. Whatever the shape of the survival curve may be, the transition to a 150-year life expectancy allows us to compensate at term the fall in the population size observed when life expectancy remained limited at 85 years. However, in the case of the extension of the age at death, the population starts to diminish during approximately one century (from 8 to 6 billion in 2135), before reaching its initial level again which is kept when the rectangularity of the survival curve is maintained.

This transitory stage of increasing distance between both trajectories at the 150-year life expectancy can also be found in the scenarios of fertility stabilization below the replacement level (Fig. 17). But, what is more remarkable here is the slowing down of the diminution of the population size, which begins to sig-

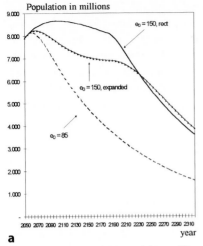
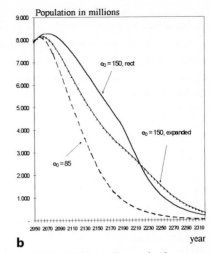

a   **b**

Fig. 17. Compared evolutions of the world's population's size (2050–2315) according to the three mortality hypotheses, in both cases in which fertility stabilizes below the replacement level: **A)** UN low hypothesis, **B)** only child.

nificantly diminish only after 150 years with 1.7 children per woman (Fig. 17 a). There is even, at the beginning, a slight population growth (from 8 to 8.7 billion in four decades) when the rectangularity of the survival curve is maintained.

The population diminution takes place far quicker, of course, in the hypothesis of the only child (Fig. 17 b), but in this case also, the slowing down is decisive. Whereas 70 years were sufficient to divide the population by two with life expectancy fixed at 85 years, when average life duration reaches 150 years, 120 years are necessary in the case of an extension of the age at death, and 150 when the rectangularity of the survival curve is maintained. When the trajectories of the two scenarios at 150-year life expectancy merge, approximately by 2215, there still remains 2.5 billion men and women instead of only 480 millions with 85-year life expectancy.

Of course, the hypothesis of a transition to 150-year life expectancy is daring but we can note how much it can make up, for centuries, for the persistence of the fertility level below the generation replacement level.

Evolution of Birth and Death Rates

The transition from an 85- to 150-year life expectancy obviously tends to reduce the crude birth and death rates, all other things being equal.

At the end of the forecast (2315), with fertility at the replacement level, whatever the calendar may be, birth and death rates are almost divided by two in the case of a transition from 85- to 150-year life expectancy (Adenda 2 and 3); they fall from 11 or 12 to 7 or 8 per thousand), which is quite logical as, in a stationary population, birth and death rates are the reverse of life expectancy. With 2.5 children per woman, the birth rate falls from 15 to 11 per thousand, but mortality is divided by more than two (from 9 to 4 per thousand). With fertility below the replacement level, the lengthening of life expectancy to 150 years brings the birth rate from 9 to 4 per thousand (1.7 children per woman) or even from 4 to 0.5 per thousand (only child). The relative variation of the death rate is smaller (respectively from 15 to 10 or from 28 to 25 per thousand) but the effect on the growth rate is equivalent.

However, these final results also follow more or less uneven trajectories, which explain those of the population size. We will study only a few examples.

With the 2.1 children per woman of the UN average hypothesis (Fig. 18), whereas birth and death rates were converging very rapidly with 85-year life expectancy, the transition to 150 years postpones this convergence for over a century. In fact, whatever the shape of the survival curve adopted may be, what is most striking is undoubtedly the contrast between the regularity of the diminution of the birth rate and the chaotic evolution of the death rate. The shock endured by the latter itself is very different, depending on whether the 150-year life expectancy is obtained when the rectangularity of the survival curve is maintained or with an extension of the age at death. This shock needs to be explained. When the rectangularity of the survival curve is maintained, the gains

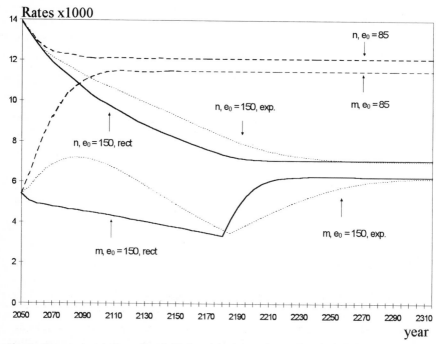

Fig. 18. Compared evolutions of crude birth and death rates (2050–2315) according to the three hypotheses of life expectancy evolution, in the case in which fertility follows the UN average hypothesis (2.1 children per woman).

realized in terms of life expectancy prevail over the population aging and the crude death rate decreases very regularly, as long as life expectancy increases.

However, as soon as life expectancy becomes fixed (our last table is acquired by 2190, Fig. 6), the population aging abruptly begins to take effect. With an extension of the age at death, on the contrary, the mortality diminution bears, at the beginning, on small numbers of very old people only. Aging is then sufficient to bring about an increase in the crude death rate. It is only around 2090 that the reverse movement starts, inducing a quicker decrease in the mortality rate than in the preceding case. Then the trend reverse once again, although less abruptly, when the last mortality table is reached.

Exactly identical types of trajectories are found in Figure 19, which illustrates the case of transition to the only child, but this time, it is naturally in the framework of an extreme difference between birth and death rates, and consequently the chaotic evolution of the latter is still more marked.

Finally, Figure 20 illustrates the dramatic fluctuations to which the two parameters of the natural increase are subjected by the transition to unimodal late fertility. It is interesting to note here that the lenghtening of life expectancy significantly reduces birth fluctuations and seems to accelerate their suppression, whereas it imparts a very different course to mortality. When the rectangularity of the survival curve is maintained, in particular, fluctuations take place only

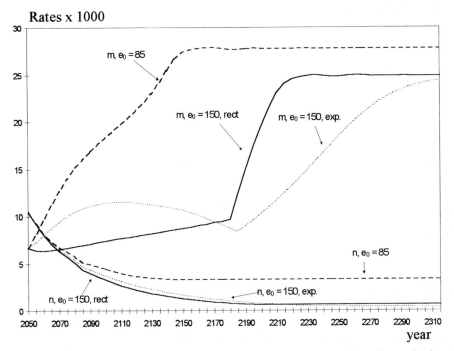

Fig. 19. Compared evolutions of the crude birth and death rates (2050–2315) according to the three hypotheses of life expectancy evolution, in the hypothesis of the transition to the only child.

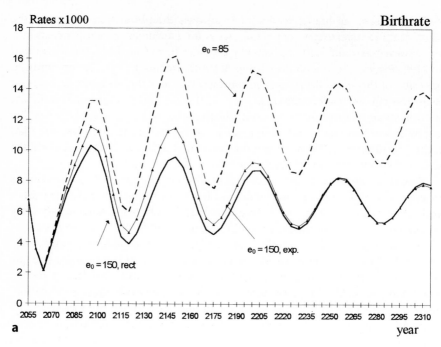

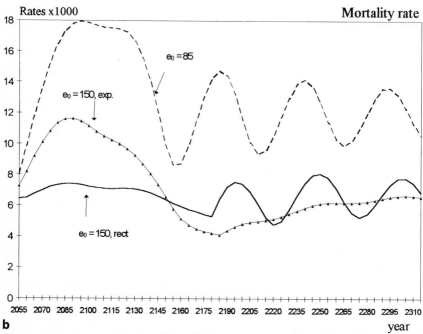

Fig. 20. Compared evolutions of the crude birth and death rates (2050–2315) according to the three hypotheses of life expectancy evolution, with unimodal late fertility at 2.1 children per woman: **a)** birth rate, **b)** mortality rate.

after the final table is reached. In the case of an extension of the age at death, there is only one very wide fluctuation left, at the very beginning, followed by a slow rise of the crude rate.

Evolution of the Age Structure

Addendum 5 sums up the evolution of the population distribution by broad age groups, over the period of the forecast, according to the 21 scenarios mentioned here.

At the end of the forecast, the two mortality hypotheses giving 150-year life expectancy provide of course, age pyramids that are very different from those reached with the UN 85-year limit (Fig. 21 a to 21 f). With the UN average fertility hypothesis, the transformation just reproduces the differences existing between the survival curves of the mortality tables (Fig. 21 a). Whereas we obtained a stabilization of the structure by broad age group over some 50 years (24 % of people below age 20 and 30 % of people above 60 as soon as 2100), not only does the transition to 150-year life expectancy considerably postpone this stabilization, which only takes place 50 years later in the case of a rectangularization of the survival curve, and even a 100 years later in the case of an extension of the age at death, but also it significantly increases the demographic aging by causing, in both cases, the proportion of people below age 20 to fall to 14 % and the proportion of people above age 60 to rise to 59 % (Addendum 5). This aging is all the more marked as it results in an unprecedented expansion of the category of people above age 100. Of the 59 % of people above 60, more than half (33 % of the total population) are older than centenarian, whatever the hypothesis on the shape of the survival curve may be.

The results at term of the adoption of the early bimodal fertility of 2.1 children per woman are not very different (Addendum 5), but they are obtained at the end of a period of longer stabilization comprising fluctuations. This should also occur in the case of bimodal late fertility, as we remain in the framework of the 2.1 children per woman hypothesis, however beyond the end of the forecast, which results in still disrupted pyramids by 2315 (Fig. 21 b). This phenomenon is pushed to the extreme in the case of unimodal late fertility. In this case also, the pyramids show many disruptions by 2315, especially with the hypothesis of a transition to 150-year life expectancy when the rectangularity of the survival curve is maintained.

However the most extravagant pyramids are naturally those provided by the two cases in which fertility falls below replacement level. This, which is already obvious with 1.7 children per woman (Fig. 21 d), is far more striking with the only-child hypothesis (Fig. 21 e). In both of these cases also, extreme profiles are provided by 150-year life expectancy with a rectangular survival curve, in which the concentration on a very high modal age is amazing. With 1.7 children per woman, the proportion of people below age 20, which was stable at 22 % with 85-year life expectancy, falls to 8 % when life expectancy shifts to 150 years, what-

a) Total fertility rate = 2,1

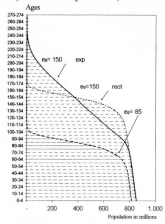

b) Fertility: two-late modes

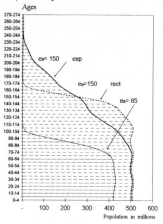

c) Fertility: one late mode

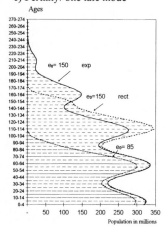

d) Total fertility rate = 1,7

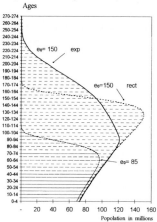

e) Total fertility rate = 1,0

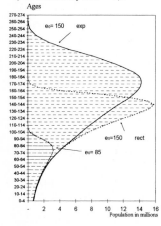

f) Total fertility rate = 2,5

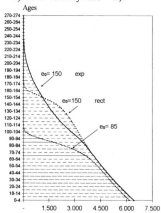

Fig. 21. Comparison of the population profiles per age reached by 2315 according to the life expectancy hypothesis, following different fertility scenarios (**a–f**).

ever the shape of the survival curve may be. At the same time, the proportion of people above age 60 shifts from 35 to more than 70 % (and even nearly 75 % in the case of an extension of the age at death). The proportion of people above age 100 reaches then 45 and 49 % depending on the shape of the survival curve (Addendum 4).

With the transition to the only child, the proportion of people below age 20, which was stabilized at 8.5 % with 85-year life expectancy, falls below 2 % when life expectancy shifts to 150 years (1.6 % if the survival curve is rectangular, and even 1 % only in the case of an extension of the age at death). At the same time, people above age 60 shift from 56 to more than 90 % (and even almost 95 % in the case of an extension of the age at death). The proportion of people above age 100 then reaches the extravagant level of 74 %, and even 84 % depending on the shape of the survival curve (Addendum 5): almost the whole population is centenarian! We are about to enter truly uncharted territory.

Of course, the maintenance of fertility at 2.5 children per woman has the exact opposite effect, as it results in younger population profiles recalling in some way our past (Fig. 21 f). In some way only, as this assumption is valid only if we admit that the notions of "young" and "old" will have evolved in the meantime. If the proportion of people below age 20 remains above 20 % whatever the mortality hypothesis may be (falling only from 25 to 20 % when we shift from 85- to 150-year life expectancy), in any way, these 20 or 25 % remain far less than what was observed in the past. And, moreover, the proportion of people above age 60, which shifts from 24 to more than 46 % when life expectancy shifts from 85 to 150 years, is greater than in any past or even present population. However, if we consider the general aspect of the age pyramid, and especially if we compare it to the results of the other forecasts, we can really call it "young". Paradoxically, it is the hypothesis of an extension of the age at death that shows the most characteristic image of this "rejuvenation", although with a proportion of centenarians superior to 22 %.

Finally, we should observe that the results obtained in the framework of the fertility hypotheses diverging from the cohort replacement level (and changing radically the population's age structure) are obtained within periods of time very similar to those observed in the UN hypothesis of 2.1 children per woman. From this perspective, it is the change in the fertility calendar that destabilizes the population's age structure rather than the change in the level, even if it results in a more classical stable structure.

References

Allard M, Lèbre V, Robine JM (1994) Les 120 ans de Jeanne Calment, doyenne de l'humanité. Paris, Le Cherche Midi

Benjamin B (1982) The span of life. Actuaries 109:319–340

Bourgeois-Pichat J (1952) Essai sur la mortalité biologique de l'homme. Population (3):381–394

Bourgeois-Pichat J (1966) Intervention. In: Comptes rendus analytiques des séances: évolution de la fécondité, ses causes et ses conséquences. Documents officiels de la conférence, vol. IV, CR2, 6–7.

Strasbourg, Conseil de l'Europe (Conférence démographique européenne, Strasbourg 30 août – 6 septembre 1966)

Bourgeois-Pichat J (1977) Le dilemme de la révolution démographique: croître ou vieillir. In: Actes du congrès sur le vieillissement organisé par l'Institut de la vie à Vichy, France. Future Age Projection Society, 260–278

Bourgeois-Pichat J (1978) Future outlook for mortality decline in the world. Pop Bull UN 11:12–41

Bourgeois-Pichat J (1987) La demografia e le altre scienze: interazioni di metodo e cotenuto, In: Sonnino E, Pinelli A, Maffioli Di Nobile A (eds) Demografia: scienza insegnamento professione. Rome, Franco Angeli, 53–67

Bourgeois-Pichat J (1988) Du XXe au XXIe siècle: l'Europe et sa population après l'an 2000. Population 43 (1): 9–44

Fries J (1982) Aging, natural death, and the compression of morbidity. New Engl J Med 303:130–135

Fries J (1989) The compression of morbidity: near or far? Milbank Quart 67:208

Guibert-Lantoine C, Monnier A (1995) La conjoncture démographique: l'Europe et les pays développés d'Outre-mer. Population 50 (14/5):1185–1210

Kannisto V (1994) Development oldest-old mortality, 1950–1990: evidence from 28 developed countries. Odense, Odense University Press

Olshansky SJ, Carnes A, Cassel C (1990) In search of Methuselah: estimating the upper limits to human longevity. Science 250:634–640

Santini A (1995) Continuità e discontinuità nel comportamento riproduttivo delle donne italiane nel dopoguerra: tendenze generali della fecondità delle coorti nelle ripartizioni tra il 1952 e 1991. Florence, Dipartimento di Statistica G. Parenti (Working papers, n53)

United Nations (1982) Projections démographiques mondiales à long terme effectuées en 1980. Pop Bull UN 14

United Nations (1992) Long-range world population projections: two centuries of population growth 1950–2150. New York, Department of international economic and Social Affairs, United Nations (Document ST/ESA/SER.A/125).

United Nations (1994) The sex and age distribution of the world populations, the 1994 revision. New York, Population Division (Document ST/ESA/SER A/144)

United Nations (1995) World population prospects, the 1994 revision. New York, United Nations Population Division (Document ST/ESA/SER A/145)

Vallin J (1989) De la mortalité endogène à l'allongement de la vie: hommage à Jean Bourgeois-Pichat In: Vallin J (ed), L'avenir de l'espérance de vie. Paris, INED/PUF (Congrès et Colloques, n 12), 23–35

Vallin J (1994) Reflexions sur l'avenir de la population mondiale, version révisée d'une communication présentée au Congrès international de la population, Montréal, septembre 1993. Paris, CEPED (Les Dossiers du CEPED, n 26)

Vallin J, Meslé F (1988) Les causes de décès en France de 1925 à 1978. Paris, INED/PUF (Travaux et Documents, Cahier, n 115)

Walford R (1982) Immunoregulation and aging, In: Preston SH (ed) Biological and social aspects of mortality and the length of life: proceedings of a seminar at Fiuggi, Italy, May 13–16, 1980. Liège, UIESP, Ordina Éditions, 259–278

Walford R (1984) Maximum life span. New York, Avon

Znata A. De Rose A (1995) Il figlio unico in Italia: frequenza e determinanti di scelta. Rome, Dipartimento di Scienze Demografische, Università degli Studi di Roma "La Sapienza" (Materiali di studi e di recerche, n 8)

Addendum 1. Evolution of the Size of the World's Population from 2050 to 2315, According to Three Mortality Hypotheses and Seven Fertility Hypotheses

Mortality and fertility hypotheses	Population (in millions)											
	2050	2075	2100	2125	2150	2175	2200	2225	2250	2275	2300	2315
UN limit life expectancy (85 years)												
UN 2.1	9.833	11.342	11.808	12.007	12.201	12.393	12.587	12.785	12.985	13.189	13.396	13.522
UN 1.7	7.918	7.882	6.764	5.680	4.822	4.089	3.466	2.939	2.492	2.113	1.791	1.622
UN 2.5	11.912	15.561	18.827	22.304	26.332	31.081	36.686	43.302	51.112	60.329	71.209	78.657
Only child	7.918	7.672	5.887	3.972	2.360	1.282	696	377	205	111	60	42
2 early modes	9.833	10.694	10.391	9.865	9.447	9.492	9.421	9.369	9.321	9.269	9.220	9.190
2 late modes	9.833	10.115	9.589	8.658	8.095	8.267	7.981	7.880	7.757	7.601	7.487	7.401
1 late mode	7.918	7.066	5.913	4.794	4.433	4.712	4.339	4.661	4.292	4.566	4.271	4.344
150-year life expectancy with extension of the age at death												
UN 2.1	9.833	11.660	13.002	14.336	15.907	17.785	19.871	21.486	22.588	23.332	23.896	24.203
UN 1.7	7.918	8.185	7.771	7.326	7.059	6.931	6.843	6.460	5.832	5.098	4.379	3.983
UN 2.5	11.912	15.895	20.220	25.409	31.940	40.249	50.559	62.016	74.721	89.141	105.889	117.329
Only child	7.918	7.975	6.890	5.600	4.478	3.611	2.957	2.243	1.530	939	534	374
2 early modes	9.833	11.011	11.587	12.181	12.928	14.083	15.440	16.381	16.907	17.159	17.270	17.308
2 late modes	9.833	10.433	10.806	11.029	11.505	12.493	13.651	14.466	14.902	15.112	15.196	15.208
1 late mode	7.918	7.368	6.923	6.429	6.424	6.985	7.431	8.013	8.048	8.351	8.217	8.323
150-year life expectancy with rectangularity of the survival curve maintained												
UN 2.1	9.833	11.966	14.014	16.004	17.989	20.014	21.631	22.234	22.695	23.164	23.646	23.940
UN 1.7	7.918	8.479	8.663	8.599	8.417	8.186	7.658	6.601	5.626	4.793	4.084	3.710
UN 2.5	11.912	16.211	21.351	27.482	34.840	43.718	53.734	64.008	75.926	90.060	106.837	118.367
Only child	7.918	8.269	7.783	6.877	5.829	4.782	3.537	2.045	1.115	607	331	230
2 early modes	9.833	11.317	12.606	13.865	14.997	16.131	16.836	16.893	17.010	17.020	17.075	17.101
2 late modes	9.833	10.740	11.846	12.757	13.607	14.461	14.902	15.072	15.220	15.159	15.199	15.226
1 late mode	7.918	7.664	7.824	7.712	7.744	8.021	7.969	8.287	8.108	8.263	8.171	8.162

Addendum 2. Evolution of the Crude Birth Rate from 2050 to 2315, According to Three Mortality Hypotheses and Seven Fertility Hypotheses

Mortality and fertility hypotheses	Birth rate (per thousand)											
	2050	2075	2100	2125	2150	2175	2200	2225	2250	2275	2300	2315
UN limit life expectancy (85 years)												
UN 2.1	13.92	12.34	12.08	12.08	12.09	12.09	12.09	12.09	12.09	12.09	12.09	12.09
UN 1.7	10.45	8.63	8.55	8.67	8.69	8.69	8.69	8.69	8.69	8.69	8.69	8.69
UN 2.5	17.21	15.87	15.54	15.49	15.48	15.48	15.48	15.48	15.48	15.48	15.48	15.48
Only child	10.45	6.26	4.43	3.57	3.30	3.30	3.30	3.30	3.30	3.30	3.30	3.30
2 early modes	13.92	10.16	11.13	11.14	11.68	11.61	11.67	11.68	11.66	11.65	11.65	11.66
2 late modes	13.92	7.94	11.45	11.49	11.16	11.59	11.42	11.29	11.51	11.37	11.40	11.39
1 late mode	10.45	6.18	13.22	7.40	16.17	7.57	15.25	8.65	13.96	9.99	12.64	13.40
150-year life expectancy with extension of the age at death												
UN 2.1	13.92	11.94	10.92	10.10	9.28	8.47	7.75	7.34	7.14	7.06	7.04	7.04
UN 1.7	10.45	8.25	7.38	6.69	5.92	5.13	4.43	4.02	3.81	3.72	3.69	3.69
UN 2.5	17.21	15.48	14.43	13.60	12.82	12.06	11.40	11.05	10.89	10.84	10.82	10.82
Only child	10.45	5.97	3.74	2.49	1.69	1.14	0.76	0.55	0.45	0.40	0.39	0.38
2 early modes	13.92	9.81	9.95	9.04	8.64	7.97	7.32	6.95	6.75	6.67	6.65	6.65
2 late modes	13.92	7.69	10.29	9.27	8.22	8.18	7.26	6.82	6.80	6.61	6.61	6.60
1 late mode	10.45	5.89	11.26	5.53	11.47	5.20	9.26	5.27	7.88	5.86	7.06	7.64
150-year life expectancy with rectangularity of the survival curve maintained												
UN 2.1	13.92	11.57	10.07	9.00	8.17	7.50	7.13	7.09	7.09	7.10	7.09	7.09
UN 1.7	10.45	7.90	6.57	5.66	4.94	4.33	3.98	3.95	3.95	3.95	3.95	3.95
UN 2.5	17.21	15.12	13.60	12.52	11.70	11.07	10.73	10.69	10.69	10.69	10.69	10.69
Only child	10.45	5.71	3.27	2.01	1.29	0.86	0.65	0.62	0.62	0.62	0.62	0.62
2 early modes	13.92	9.50	9.11	7.92	7.45	6.97	6.76	6.75	6.72	6.72	6.72	6.72
2 late modes	13.92	7.46	9.44	8.08	7.03	7.17	6.78	6.63	6.75	6.67	6.69	6.67
1 late mode	10.45	5.63	9.94	4.60	9.57	4.55	8.71	5.10	7.86	5.91	7.11	7.78

Addendum 3. Evolution of the Crude Death Rate from 2050 to 2315, According to Three Mortality Hypotheses and Seven Fertility Hypotheses

Mortality and fertility hypotheses	Mortality rate (per thousand)											
	2050	2075	2100	2125	2150	2175	2200	2225	2250	2275	2300	2315
UN limit life expectancy (85 years)												
UN 2.1	5.42	9.84	11.32	11.44	11.48	11.47	11.46	11.46	11.46	11.46	11.46	11.46
UN 1.7	6.68	13.51	15.76	15.25	15.38	15.33	15.29	15.28	15.28	15.29	15.29	15.29
UN 2.5	4.52	7.54	8.58	8.84	8.85	8.85	8.85	8.85	8.85	8.85	8.85	8.85
Only child	6.68	13.96	18.35	22.00	27.37	27.76	27.78	27.75	27.73	27.72	27.73	27.73
2 early modes	5.42	10.50	12.89	13.62	11.95	11.74	12.02	11.83	11.88	11.86	11.87	11.87
2 late modes	5.42	11.16	13.97	15.18	11.23	12.19	12.10	12.09	12.13	12.04	12.16	12.08
1 late mode	6.68	15.20	17.86	17.22	9.92	13.04	11.56	11.76	12.65	10.94	13.17	10.58
150-year life expectancy with extension of the age at death												
UN 2.1	5.42	7.06	7.03	6.09	4.93	3.87	4.08	4.95	5.63	6.02	6.18	6.21
UN 1.7	6.68	9.68	10.01	8.57	6.92	5.49	5.93	7.46	8.77	9.60	9.98	10.06
UN 2.5	4.52	5.39	5.21	4.49	3.60	2.78	2.86	3.37	3.72	3.91	3.98	3.99
Only child	6.68	9.99	11.41	11.35	10.65	9.08	10.26	14.09	18.38	22.00	23.93	24.38
2 early modes	5.42	7.51	7.91	7.13	5.68	4.17	4.39	5.30	5.95	6.33	6.49	6.53
2 late modes	5.42	7.97	8.49	7.82	6.01	4.14	4.48	5.35	5.94	6.34	6.52	6.57
1 late mode	6.68	10.83	11.14	9.67	6.55	4.32	4.81	5.30	6.14	6.21	6.63	6.58
150-year life expectancy with rectangularity of the survival curve maintained												
UN 2.1	5.42	4.75	4.45	4.12	3.77	3.42	5.67	6.25	6.28	6.27	6.27	6.27
UN 1.7	6.68	6.49	6.59	6.36	5.99	5.51	9.25	10.27	10.36	10.36	10.35	10.35
UN 2.5	4.52	3.63	3.18	2.82	2.48	2.18	3.51	3.85	3.87	3.86	3.86	3.86
Only child	6.68	6.69	7.39	8.03	8.74	9.47	19.56	24.76	24.81	24.83	24.85	24.85
2 early modes	5.42	5.04	4.95	4.75	4.50	4.17	6.74	6.25	6.67	6.61	6.62	6.63
2 late modes	5.42	5.32	5.26	5.16	4.95	4.58	7.11	5.44	7.16	6.58	6.53	6.67
1 late mode	6.68	7.21	7.24	7.11	6.40	5.44	7.41	4.99	8.10	5.27	7.76	6.86

Addendum 4. Evolution of the Crude Rate of Natural Increase, from 2050 to 2315, According to Three Mortality Hypotheses and Seven Fertility Hypotheses

Mortality and fertility hypotheses	Natural increase rate (per thousand)											
	2050	2075	2100	2125	2150	2175	2200	2225	2250	2275	2300	2315
UN limit life expectancy (85 years)												
UN 2.1	8.50	2.49	0.76	0.64	0.61	0.62	0.62	0.62	0.62	0.62	0.62	0.62
UN 1.7	3.77	-4.88	-7.21	-6.58	-6.69	-6.63	-6.60	-6.59	-6.60	-6.60	-6.60	-6.60
UN 2.5	12.69	8.33	6.96	6.65	6.64	6.63	6.63	6.63	6.63	6.63	6.63	6.63
Only child	3.77	-7.70	-13.93	-18.43	-24.07	-24.46	-24.48	-24.45	-24.43	-24.42	-24.43	-24.42
2 early modes	8.50	-0.34	-1.77	-2.48	-0.26	-0.12	-0.34	-0.15	-0.23	-0.20	-0.22	-0.22
2 late modes	8.50	-3.22	-2.52	-3.69	-0.07	-0.60	-0.67	-0.80	-0.62	-0.67	-0.77	-0.69
1 late mode	3.77	-9.01	-4.64	-9.82	-6.25	-5.47	3.69	-3.11	1.32	-0.96	-0.53	2.82
150-year life expectancy with extension of the age at death												
UN 2.1	8.50	4.88	3.89	4.01	4.36	4.60	3.67	2.39	1.51	1.04	0.86	0.83
UN 1.7	3.77	-1.43	-2.63	-1.89	-1.00	-0.37	-1.50	-3.44	-4.96	-5.89	-6.28	-6.37
UN 2.5	12.69	10.09	9.22	9.11	9.22	9.28	8.54	7.68	7.17	6.93	6.84	6.83
Only child	3.77	-4.02	-7.68	-8.86	-8.96	-7.94	-9.50	-13.54	-17.94	-21.60	-23.55	-23.99
2 early modes	8.50	2.30	2.04	1.91	2.97	3.80	2.94	1.65	0.80	0.35	0.16	0.12
2 late modes	8.50	-0.29	1.81	1.45	2.21	4.04	2.78	1.47	0.85	0.27	0.09	0.03
1 late mode	3.77	-4.94	0.12	-4.14	4.92	0.88	4.44	-0.03	1.73	-0.34	0.43	1.06
150-year life expectancy with rectangularity of the survival curve maintained												
UN 2.1	8.50	6.82	5.62	4.88	4.40	4.08	1.47	0.85	0.81	0.82	0.82	0.82
UN 1.7	3.77	1.41	-0.02	-0.70	-1.05	-1.18	-5.27	-6.32	-6.42	-6.41	-6.40	-6.40
UN 2.5	12.69	11.49	10.42	9.69	9.22	8.89	7.21	6.84	6.83	6.83	6.83	6.83
Only child	3.77	-0.97	-4.12	-6.03	-7.45	-8.61	-18.91	-24.14	-24.19	-24.20	-24.23	-24.23
2 early modes	8.50	4.46	4.17	3.16	2.94	2.80	0.02	0.50	0.04	0.12	0.10	0.09
2 late modes	8.50	2.14	4.18	2.92	2.08	2.59	-0.33	1.19	-0.41	0.09	0.16	0.00
1 late mode	3.77	-1.58	2.70	-2.51	3.18	-0.89	1.30	0.11	-0.25	0.64	-0.65	0.92

Addendum 5. Evolution (2050–2315) of the World Population's Distribution by Age Groups, According to Three Mortality Hypotheses and Seven Fertility Hypotheses

Mortality and fertility hypotheses	Age	Population ratio (percentage)											
		2050	2075	2100	2125	2150	2175	2200	2225	2250	2275	2300	2315
UN limit life expectancy (85 years)													
UN 2.1	0–19	27.7	24.5	24.0	23.9	23.9	23.9	23.9	23.9	23.9	23.9	23.9	23.9
	20–59	52.3	47.6	46.4	46.4	46.4	46.4	46.4	46.4	46.4	46.4	46.4	46.4
	60+	20.0	27.9	29.6	29.7	29.7	29.7	29.7	29.7	29.7	29.7	29.7	29.7
	Total	100.0	100.0	100.0	100.0	100.0	100.0	100.0	100.0	100.0	100.0	100.0	100.0
UN 1.7	0–19	25.2	21.9	21.7	21.9	21.8	21.8	21.8	21.8	21.8	21.8	21.8	21.8
	20–59	51.2	43.4	42.2	42.7	42.8	42.8	42.8	42.8	42.8	42.8	42.8	42.8
	60+	23.6	34.7	36.1	35.4	35.4	35.4	35.3	35.3	35.4	35.4	35.4	35.4
	Total	100.0	100.0	100.0	100.0	100.0	100.0	100.0	100.0	100.0	100.0	100.0	100.0
UN 2.5	0–19	32.9	29.7	29.0	28.9	28.9	28.9	28.9	28.9	28.9	28.9	28.9	28.9
	20–59	50.5	48.0	47.1	47.0	46.9	46.9	46.9	46.9	46.9	46.9	46.9	46.9
	60+	16.6	22.3	23.9	24.2	24.2	24.2	24.2	24.2	24.2	24.2	24.2	24.2
	Total	100.0	100.0	100.0	100.0	100.0	100.0	100.0	100.0	100.0	100.0	100.0	100.0
Only child	0–19	21.2	16.3	11.5	9.3	8.5	8.5	8.5	8.5	8.5	8.5	8.5	8.5
	20–59	53.9	46.5	45.2	39.5	36.0	35.9	35.9	35.9	35.9	35.9	35.9	35.9
	60+	24.8	37.1	43.2	51.2	55.5	55.6	55.6	55.6	55.6	55.6	55.6	55.6
	Total	100.0	100.0	100.0	100.0	100.0	100.0	100.0	100.0	100.0	100.0	100.0	100.0
2 early modes	0–19	27.7	20.0	21.4	22.9	23.4	23.3	23.2	23.2	23.2	23.3	23.3	23.2
	20–59	52.3	50.5	45.0	43.3	47.0	45.9	46.3	46.3	46.3	46.3	46.3	46.3
	60+	20.0	29.6	33.7	33.8	29.6	30.8	30.5	30.4	30.5	30.5	30.5	30.5
	Total	100.0	100.0	100.0	100.0	100.0	100.0	100.0	100.0	100.0	100.0	100.0	100.0

Mortality and fertility hypotheses	Age	Population ratio (percentage)											
		2050	2075	2100	2125	2150	2175	2200	2225	2250	2275	2300	2315
2 late modes	0–19	27.7	15.4	22.3	21.5	23.4	23.0	22.5	23.1	22.8	22.8	23.0	22.9
	20–59	52.3	53.4	41.2	42.3	49.5	45.0	46.4	46.4	46.0	46.3	46.2	46.2
	60+	20.0	31.2	36.5	36.2	27.1	32.0	31.0	30.5	31.2	30.9	30.9	31.0
	Total	100.0	100.0	100.0	100.0	100.0	100.0	100.0	100.0	100.0	100.0	100.0	100.0
1 late mode	0–19	21.2	9.2	22.5	18.4	25.8	22.3	23.0	24.5	21.1	25.7	20.4	25.3
	20–59	53.9	50.5	34.5	44.7	48.7	42.0	51.2	40.8	51.9	41.2	51.0	46.1
	60+	24.8	40.3	43.1	36.8	25.5	35.6	25.8	34.6	27.0	33.2	28.6	28.7
	Total	100.0	100.0	100.0	100.0	100.0	100.0	100.0	100.0	100.0	100.0	100.0	100.0
150-year life expectancy with extension of the age at death													
UN 2.1	0–19	27.7	23.9	21.9	20.2	18.6	16.9	15.5	14.6	14.2	14.0	14.0	14.0
	20–59	52.3	46.4	42.4	39.3	36.2	33.1	30.2	28.5	27.7	27.3	27.2	27.2
	60–99	19.9	29.2	33.0	33.3	32.4	30.5	28.3	26.8	26.0	25.7	25.6	25.6
	100+	0.1	0.5	2.7	7.2	12.9	19.5	26.0	30.1	32.2	33.0	33.2	33.2
	Total	100.0	100.0	100.0	100.0	100.0	100.0	100.0	100.0	100.0	100.0	100.0	100.0
UN 1.7	0–19	21.2	17.9	16.1	14.5	12.8	11.1	9.6	8.6	8.2	7.9	7.9	7.9
	20–59	53.9	43.7	38.5	34.9	31.0	26.9	23.2	21.0	19.8	19.3	19.1	19.1
	60–99	24.7	37.7	41.0	39.3	37.0	33.1	29.0	26.2	24.8	24.1	23.9	23.9
	100+	0.1	0.7	4.4	11.3	19.2	28.9	38.2	44.2	47.3	48.6	49.1	49.1
	Total	100.0	100.0	100.0	100.0	100.0	100.0	100.0	100.0	100.0	100.0	100.0	100.0
UN 2.5	0–19	32.9	29.1	27.1	25.5	24.1	22.7	21.4	20.7	20.4	20.3	20.2	20.2
	20–59	50.5	47.1	44.2	41.7	39.3	37.0	35.0	33.8	33.3	33.1	33.0	33.0
	60–99	16.5	23.4	26.9	28.0	27.7	26.9	25.8	25.0	24.6	24.5	24.4	24.4
	100+	0.1	0.3	1.9	4.8	8.9	13.4	17.8	20.5	21.7	22.2	22.3	22.3
	Total	100.0	100.0	100.0	100.0	100.0	100.0	100.0	100.0	100.0	100.0	100.0	100.0
	0–19	21.2	15.7	9.9	6.7	4.5	3.1	2.0	1.5	1.2	1.0	1.0	1.0
	20–59	53.9	44.9	38.9	28.4	19.3	13.0	8.7	6.3	5.0	4.4	4.3	4.2

Population ratio (percentage)

Mortality and fertility hypotheses	Age	2050	2075	2100	2125	2150	2175	2200	2225	2250	2275	2300	2315
Only child	60–99	24.7	38.7	46.2	50.2	45.9	32.5	22.1	15.9	12.7	11.3	10.8	10.7
	100+	0.1	0.7	4.9	14.8	30.3	51.4	67.2	76.4	81.1	83.2	83.9	84.1
	Total	100.0	100.0	100.0	100.0	100.0	100.0	100.0	100.0	100.0	100.0	100.0	100.0
2 early modes	0–19	27.7	19.4	19.3	18.8	17.5	16.1	14.7	13.9	13.5	13.3	13.3	13.3
	20–59	52.3	49.1	40.6	35.6	35.1	31.9	29.3	27.7	26.9	26.6	26.5	26.5
	60–99	19.9	30.9	37.0	37.2	31.6	30.5	28.4	26.7	26.0	25.6	25.6	25.6
	100+	0.1	0.5	3.1	8.4	15.8	21.4	27.5	31.7	33.6	34.4	34.7	34.7
	Total	100.0	100.0	100.0	100.0	100.0	100.0	100.0	100.0	100.0	100.0	100.0	100.0
2 late modes	0–19	27.7	15.0	20.1	17.5	17.2	16.3	14.4	14.0	13.4	13.2	13.3	13.2
	20–59	52.3	51.9	36.9	33.9	36.3	31.5	29.2	27.8	26.8	26.5	26.4	26.4
	60–99	19.9	32.7	39.7	39.2	28.6	31.3	28.4	26.5	26.1	25.6	25.6	25.6
	100+	0.1	0.5	3.3	9.3	17.8	20.9	27.9	31.7	33.7	34.6	34.7	34.8
	Total	100.0	100.0	100.0	100.0	100.0	100.0	100.0	100.0	100.0	100.0	100.0	100.0
1 late mode	0–19	21.2	8.8	19.4	14.0	18.3	15.6	14.0	15.0	11.9	15.0	11.4	14.3
	20–59	53.9	48.6	29.7	33.8	34.5	29.2	31.2	25.0	29.4	24.1	28.6	26.2
	60–99	24.7	41.9	46.0	39.3	26.2	34.5	25.3	28.9	24.9	26.2	25.6	23.5
	100+	0.1	0.7	4.9	12.9	21.1	20.7	29.6	31.1	33.9	34.6	34.3	36.1
	Total	100.0	100.0	100.0	100.0	100.0	100.0	100.0	100.0	100.0	100.0	100.0	100.0
150-year life expectancy with rectangularity of the survival curve maintained													
UN 2.1	0–19	27.7	23.3	20.2	18.0	16.4	15.0	14.2	14.1	14.1	14.1	14.1	14.1
	20–59	52.3	45.3	39.4	35.2	31.9	29.2	27.6	27.4	27.4	27.4	27.4	27.4
	60–99	19.9	30.7	34.8	33.3	30.8	28.2	26.6	26.4	26.4	26.4	26.4	26.4
	100+	0.1	0.7	5.6	13.5	21.0	27.6	31.6	32.1	32.1	32.1	32.1	32.1
	Total	100.0	100.0	100.0	100.0	100.0	100.0	100.0	100.0	100.0	100.0	100.0	100.0

Mortality and fertility hypotheses	Age	Population ratio (percentage)											
		2050	2075	2100	2125	2150	2175	2200	2225	2250	2275	2300	2315
UN 1.7	0–19	21.2	17.3	14.4	12.3	10.7	9.4	8.5	8.4	8.4	8.4	8.4	8.4
	20–59	53.9	42.3	34.6	29.7	25.9	22.7	20.7	20.4	20.4	20.4	20.4	20.4
	60–99	24.7	39.4	42.4	37.7	33.6	29.3	26.6	26.2	26.3	26.3	26.3	26.3
	100+	0.1	1.0	8.6	20.3	29.8	38.7	44.2	44.9	44.9	44.9	44.9	44.9
	Total	100.0	100.0	100.0	100.0	100.0	100.0	100.0	100.0	100.0	100.0	100.0	100.0
UN 2.5	0–19	32.9	28.5	25.6	23.5	22.0	20.8	20.1	20.0	20.0	20.0	20.0	20.0
	20–59	50.5	46.3	41.9	38.5	35.9	33.9	32.7	32.6	32.6	32.6	32.6	32.6
	60–99	16.5	24.6	28.7	28.6	27.2	25.7	24.8	24.7	24.7	24.7	24.7	24.7
	100+	0.1	0.5	3.8	9.4	14.9	19.6	22.4	22.7	22.7	22.7	22.7	22.7
	Total	100.0	100.0	100.0	100.0	100.0	100.0	100.0	100.0	100.0	100.0	100.0	100.0
Only child	0–19	21.2	15.2	8.7	5.4	3.5	2.3	1.7	1.6	1.6	1.6	1.6	1.6
	20–59	53.9	43.4	34.5	23.1	14.8	9.8	7.2	6.8	6.8	6.8	6.8	6.8
	60–99	24.7	40.4	47.2	46.1	38.7	25.9	19.1	17.9	17.9	17.9	17.9	17.9
	Total	100.0	100.0	100.0	100.0	100.0	100.0	100.0	100.0	100.0	100.0	100.0	100.0
2 early modes	0–19	27.7	18.9	17.7	16.5	15.0	14.1	13.5	13.5	13.5	13.4	13.4	13.4
	20–59	52.3	47.9	37.4	31.3	30.2	27.8	26.9	26.8	26.7	26.8	26.7	26.7
	60–99	19.9	32.4	38.7	36.6	29.5	27.8	26.8	26.5	26.5	26.5	26.5	26.5
	100+	0.1	0.8	6.2	15.6	25.2	30.3	32.9	33.2	33.4	33.3	33.3	33.3
	Total	100.0	100.0	100.0	100.0	100.0	100.0	100.0	100.0	100.0	100.0	100.0	100.0
2 late modes	0–19	27.7	14.5	18.5	15.3	14.7	14.3	13.3	13.6	13.3	13.4	13.4	13.4
	20–59	52.3	50.5	33.8	29.5	31.0	27.5	27.1	27.0	26.5	26.7	26.7	26.6
	60–99	19.9	34.2	41.1	38.2	26.5	28.3	27.1	26.4	26.5	26.5	26.6	26.5
	100+	0.1	0.8	6.6	17.0	27.8	29.9	32.5	33.0	33.7	33.4	33.3	33.5
	Total	100.0	100.0	100.0	100.0	100.0	100.0	100.0	100.0	100.0	100.0	100.0	100.0
1 late mode	0–19	11.8	15.2	11.5	14.6	24.1	29.1	24.4	28.8	26.6	27.0	26.6	24.6
	60–99	24.7	43.6	46.9	37.4	23.8	31.3	24.5	28.5	25.5	33.4	33.1	34.2
	100+	0.1	1.1	9.5	22.6	32.4	29.6	33.4	32.8	33.5			
	Total	100.0	100.0	100.0	100.0	100.0	100.0	100.0	100.0	100.0	100.0	100.0	100.0

Is There A Biological Limit To The Human Life Span?

T. B. L. Kirkwood[*]

Introduction

Human life span is a quantitative trait which, like height, shows a statistical distribution within the population. A part of the variation in life span is due purely to chance (e.g., road traffic accidents); another part is due to social and behavioral factors such as life style, nutrition, medical care; a third part is due to genetics. The actual distribution of life span varies between populations and in many cases the distribution has changed profoundly over the last century. In general, the occurrence of death at younger ages has reduced, due to improved sanitation, nutrition and public health measures such as the development of vaccines and antibiotics. In consequence, the occurrence of death at older ages has proportionately increased, resulting in general increases in life expectancy. An increasing fraction of the population, especially but not solely in developed countries, is living to ages where life span appears to be limited by intrinsic failures of the biological organism associated with senescence.

As human life expectancy increases it is natural to ask if there is a biological limit to human life span which will impede further developments in longevity. Will the current maximum of 121 years, established this year by Mme. Jeanne Calment, be broken time and again? To answer this question, it is essential to understand the genetic contribution to the life span distribution and to consider how genetic and non-genetic factors may interact. One preliminary observation should be noted: maximum life span is an extreme value statistic. Like the world record for running the 1500 metres, it can move in only one direction. This is true even if the underlying statistical distribution does not change. Therefore, it can be predicted with confidence that Mme. Calment's record will be exceeded. The question is therefore not whether this will happen, but how often and by how much.

[*] Biological Gerontology Group, University of Manchester, 3.239 Stopford Building, Oxford Road, Manchester M 13 9PT, UK

J.-M. Robine et al. (Eds.)
Longevity: To the Limits and Beyond
© Springer-Verlag Berlin Heidelberg New York 1997

Genetics of Life Span

To understand the role of genetics in ageing it is essential to consider the evolutionary forces that affect the life history (Kirkwood 1985; Finch 1990; Kirkwood and Rose 1991; Partridge and Barton 1993). In the simplest view, it may be imagined that natural selection has acted to produce a gene or set of genes that act in a clock-like manner directly to regulate duration of life, just as genes regulate growth to a final, adult size. The idea of such "ageing genes" has widespread appeal but does not make good evolutionary sense (Medawar 1952; Kirkwood and Cremer 1982). In the first place, there is little evidence that the process of ageing really acts in nature in the way this theory suggests. Most animals in their natural environment die well before senescent changes become apparent and certainly before the end stages of senescence set in. This limits the scope of natural selection to direct the evolution of ageing genes, when their expression under natural circumstances is a rarity. A second objection is that the required selection force must apparently work in the wrong direction. Senescence is detrimental to the individual organism and natural selection should oppose, not favour, it. The suggestion has been made that ageing is necessary as a form of population control (to prevent old organisms from taking resources and living space needed by the young), but this argument is seriously compromised by the fact that it relies on the problematic concept of "group selection" and there is little evidence to support it. [It should be noted that the mass "senescence" of species like Pacific salmon occurs within the context of a life history that involves semelparous (once-only) reproduction; a different pattern of selection forces arises in such cases, compared with that which applies in species showing iteroparous (repeated) reproduction (Kirkwood 1985).]

Instead of ageing genes, evolution theory suggests that the genetic basis of ageing is due to two factors: 1) natural selection is little concerned with events that occur late in the life span, because these have negligible effects on genetic fitness; and 2) the acquisition of a greater longevity is likely to involve some direct cost. Specifically, the disposable soma theory (Kirkwood 1977, 1981; Kirkwood and Holliday 1979) predicts that ageing is due to evolved limitation in the efficiency of fundamental processes of somatic maintenance and repair.

The disposable soma theory is based on recognition that, because of hazards such as predation, starvation and disease, expectation of life in the wild environment is finite, even in the absence of intrinsic deterioration. When this observation is coupled with the fact that maintenance processes (DNA repair, antioxidant enzymes, stress responses, etc.) all require energy and other resources to fuel them, it is found that the optimum course is to invest only as many resources in maintenance of somatic cells and tissues as are necessary for the individual to remain in good condition during the natural expectation of life in the wild environment. A greater level of maintenance is actually disadvantageous because it eats into resources that, in terms of natural selection, are better used for reproduction.

The upshot of the disposable soma theory is that for a species in its natural ("wild") environment most individuals will die from accidental causes before intrinsic deterioration becomes important. However, when the level of hazard is reduced, as in the case of protected animal populations or, for our own species, following a period of rapid social development, many individuals live long enough to encounter the deleterious effects of accumulated somatic damage, i.e., ageing.

It may be noted that, in addition to the accumulation of somatic damage, other genetic factors may be involved in the evolution of ageing. These include germ-line mutations causing deleterious effects late in life that simply fail to get removed due to the weak action of selection at older ages (Medawar 1952), and pleiotropic effects involving trade-offs between gene actions that are beneficial early in life but cause harm at later ages (Williams 1957). Such factors will not be discussed in detail in what follows since the gene actions are of a general type; however, it should be readily apparent that they can be accommodated within the same framework (Kirkwood and Rose 1991).

Implications for the Distribution of Life Span

The Ageing Process is Stochastic

The disposable soma theory implies that the primary processes contributing to biological ageing are the accumulations over time of unrepaired somatic defects, many of which arise as by-products of intrinsic metabolism. The individual events contributing to these accumulations are random, or stochastic, but the rates of accumulation are regulated genetically through the levels of the various maintenance functions. For example, by varying the level of a particular DNA repair activity, the rate of accumulation of the corresponding DNA damage is affected. Although the individual events are stochastic, the numbers of events involved are often large enough that the apparent random variation is not obvious. An analogous situation occurs with radiocarbon dating where the decay of individual atomic nuclei is random but the behaviour of the population of atoms in a particular sample is highly reproducible. Stochastic effects will be most apparent where the numbers of initiating events for an age-related change are small. This is particularly true for DNA damage affecting cell proliferation. A few mutations in a single cell may result in abnormal proliferation (hyperplasia) or malignant neoplasia.

The Ageing Process is Polygenic

The disposable soma theory firmly predicts a polygenic basis for the distribution of life spans (Kirkwood and Franceschi 1992). This is because the optimality principle that underpins the theory applies with equal force to each of the differ-

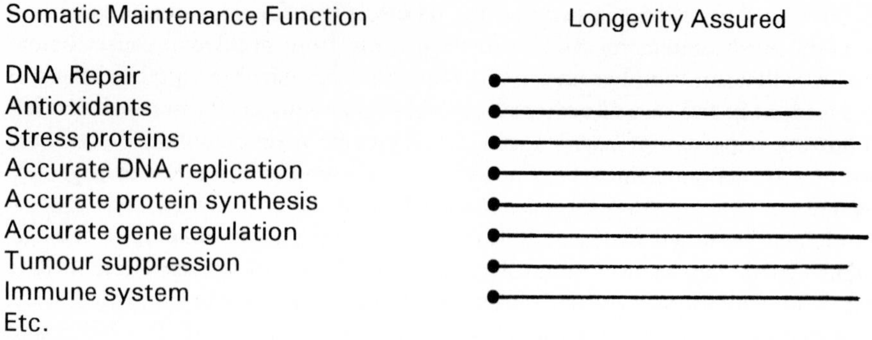

Somatic Maintenance Function   Longevity Assured

DNA Repair
Antioxidants
Stress proteins
Accurate DNA replication
Accurate protein synthesis
Accurate gene regulation
Tumour suppression
Immune system
Etc.

Fig. 1. The diagram illustrates how polygenic control of life span is effected, as predicted by the disposable soma theory of ageing. Selection acts in a similar way on the different genes regulating specific maintenance functions. The precise setting of each function determines the period of "longevity assured", with high settings giving long periods and low settings short periods. For each function, the setting is optimised so as to balance 1) the advantage of a longer period assured versus 2) the disadvantage of a greater metabolic cost. Because the chief factor determining the optimum balance is the same for each function, i.e., the level of environmental mortality (see text), it is predicted that on average the periods of longevity will be similar. However, some variance within the population is to be expected because the optimisation process is not exact. This is why the lengths of the individual lines may vary, as shown, being longer in one individual for one function than another, but perhaps being reversed in their relative lengths in a different individual (adapted from Kirkwood and Franceschi 1992).

ent maintenance and repair systems that protect against accumulation of somatic damage. There is a large number of such mechanisms (see Fig. 1). Evolutionary considerations also lead us to expect that, on average, the longevity assured by individual maintenance systems will be similar. This is because if the setting of any one mechanism is reduced so that failure occurs mainly from this cause alone, then selection will tend to act to increase this setting. Conversely, any mechanism that is set too high, relative to the others, will incur disproportionate costs and selection will reduce it.

In spite of this general harmonisation predicted for the different mechanisms contributing to ageing, some variation within populations is to be expected. The optimisation process is not absolutely precise; as a character nears the optimum, the intensity of the selection force grows smaller. This leads to the prediction that within the population there may be some variance about the mean in the actual settings of the different maintenance functions. The longest-lived individuals, such as Mme. Calment, are predicted to be those who, by good fortune, have high settings for all of the most important maintenance systems. Such a model is compatible with the data on heritability of human life span (Schächter et al. 1993).

An important corollary to the predicted polygenic basis of ageing is that the mechanisms participate in a synergistic network (Kirkwood and Franceschi 1992). This brings a much-needed coherence to the study of diverse mechanisms but it also highlights the theoretical and experimental difficulties. Recent theoretical work on a network model of cellular ageing has studied the interactions of three important processes: oxidative damage by free radicals and the protective role of antioxidants, production of aberrant proteins and removal by proteolysis, and formation and turnover of defective mitochondria with mutations to the mitochondrial genome (Kowald and Kirkwood 1996). Simulations of the network model indicated that interactions between these processes can be of major importance. It was demonstrated that the combined model had much greater explanatory and predictive power than independent sub-models describing the individual processes on their own. This has important implications for experimental study of the mechanisms of ageing which tends to focus on one mechanism at a time.

The Ageing Process is Malleable

The ageing process can obviously be altered because, over the course of evolution, species have acquired quite different distributions of life span. Humans are the longest-lived mammalian species with life spans twice as long as our near evolutionary relatives amongst the primates. The factors governing evolutionary divergence of life spans can be readily understood in the context of the disposable soma theory, namely, through ecological factors that affect the level of environmental risk. For a species occupying a high risk environment, the optimum investment in maintenance will be relatively low, with a greater emphasis on fecundity. This will result in more rapid senescence. Conversely, an adaptation that reduces the level of environmental risk can be expected to result in an increased optimum level of maintenance. Thus, the considerable longevity of flying birds and bats can be easily explained, as can the evolution of increased longevity in association with greater brain size, social living, etc., in the hominid ancestral lineage.

The idea that ageing results from accumulation of somatic damage indicates that, in principle, aspects of the ageing process may be malleable by modifying the exposure to damaging agents and/or enhancing maintenance functions. The next section briefly reviews some of the ways in which life span has been increased in animal models.

Routes to Increase Biological Life Span

Artificial Selection

In the fruitfly Drosophila melanogaster, life span has been increased by both indirect and direct selection procedures. Indirect selection has been carried out through selecting for late reproductive ability (Rose 1984; Luckinbill and Clare 1985). "Old" lines were selected from flies prevented from mating until advanced ages, whereas "young" lines were mated early. Progressively increasing adult longevity was observed in the "old" lines over several generations of selection and there is preliminary evidence indicating that the increased life span is associated with increased stress resistance (Service 1987). Recently, an ingenious direct selection procedure was developed where temperature effects on fruitfly life span were exploited (Zwaan et al. 1995). Parallel stocks were maintained both at 15 °C and 29 °C. At 29°C life span was completed in less than 60 days. Meanwhile, at 15 °C the flies remained fertile past 60 days and these could be used to breed progeny from the kin of the longest living groups. An increase in life span was produced after only a few rounds of selection and was associated with a reduction in fecundity, fulfilling one of the predictions of evolution theory. It remains to be shown whether or not these changes in life span involved alteration to the maintenance systems.

Mutational Screening

The nematode Caenorhabditis elegans is becoming established as an attractive model for investigating genetic effects on life span. A number of mutants have been identified which have altered life span (Johnson 1987; Johnson and Lithgow 1992). The first of these, age-1, shows about a 70 % increase in life span and displays a phenotype with increased resistance to stresses including heat and UV irradiation. Age-1 mutants also appear to accumulate mitochondrial mutations at a slower rate than wild-type controls. Increased thermotolerance appears to be a general character of extended-life span mutants in this species (Lithgow et al. 1995).

Transgenics

Transgenic animal models offer a powerful tool to investigate the role of maintenance systems in regulating life span. In principle, a model can be constructed in which a particular function is altered in either an upward or downward direction. The value of this approach was clearly demonstrated with transgenic Drosophila melanogaster that were made to overexpress the antioxidant enzymes superoxide dismutase (SOD) and catalase (Orr and Sohal 1994). A resulting increase in life span was observed, providing direct support for the role of the antioxidant system in longevity assurance. This model also demonstrates the need for taking account of interactions within the network. SOD and catalase operate in partnership to render the superoxide radical harmless, with SOD converting superoxide to hydrogen peroxide (itself a potent oxidising agent) and catalase converting hydrogen peroxide to water. Transgenics increasing SOD or catalase alone produced no significant increase in life span.

Nutrition

It has been long established that nutrition can affect life span, notably in the caloric restriction model of laboratory rodents (Weindruch and Walford 1988; Masoro 1992). This effect is not yet fully explained but there is growing evidence that it involves systemic upregulation of endogenous stress response mechanisms. Similar effects are observed in some invertebrate models and are currently under investigation in primates.

The idea that nutritional supplements, e.g., antioxidants, can influence life span has some support and merits further study. It is important that such studies control for confounding variables such as food intake; for example if a nutritional supplement reduces food palatibility, it may result in reduced calorie intake.

Discussion

From the biological point of view, there would appear not to be a fixed limit to human life span, even though for practical purposes the longevity of our species is rather well-defined and may prove resistant to change. The focus on maximum life span is potentially misleading, given that the longest individual life span is an extreme value statistic from a sample of some billions. The importance of such extremes depends very much on the nature and form of the underlying statistical distribution, and the significance of apparent "outliers" must be carefully weighed (Smith 1994). The human population is genetically heterogenous and the overall distribution of life span is more properly regarded as a mixture of distributions for various subpopulations. Centenarians and super-centenarians may comprise a genetic subgroup that is different but not distinct from the rest of the population.

If this point of view is correct, then the final answers to the question, "Is there a biological limit to the human life span?" can only come from further research on the genetics of life span and on the mechanisms contributing to the ageing process. With this knowledge will come the detailed understanding that will permit us to consider the feasibility of extending the quality of life in old age. Growing numbers of individuals are living to the limit of their biological potential, creating new opportunities for longevity records to be broken. The record-breakers are important and they provide a stimulus and motivation for research. However, the major focus for research must be to address the main body of the life span distribution, i.e., the general population, and to improve knowledge of the causes of age-associated morbidity and impaired quality of life.

References

Finch CE (1990) Longevity, senescence and the genome. Chicago University Press, Chicago

Johnson TE (1987) Aging can be genetically dissected into component processes using long-lived lines of Caenorhabditis elegans. Proc Natl Acad Sci USA 84:3777–3781

Johnson TE, Lithgow GJ (1992) The search for the genetic basis of aging: the identification of geronto-genes in the nematode Caenorhabditis elegans. J Am Geriatr Soc 40:936–945

Kirkwood TBL (1977) Evolution of ageing. Nature 270:301–304

Kirkwood TBL (1981) Evolution of repair: survival versus reproduction. In: Townsend CR, Calow P (eds) Physiological ecology: an evolutionary approach to resource use. Blackwell Scientific, Oxford, pp 165–189

Kirkwood TBL (1985) Comparative and evolutionary aspects of longevity. In: Finch CE, Schneider EL (eds) Handbook of the biology of aging, 2nd edn. Van Nostrand Rheinhold, New York, pp 27–44

Kirkwood TBL, Cremer T (1982) Cytogerontology since 1881: a reappraisal of August Weismann and a review of modern progress. Human Genet 60:101–121

Kirkwood TBL, Franceschi C (1992) Is aging as complex as it would appear? Ann NY Acad Sci 663:412–417

Kirkwood TBL, Holliday R (1979) The evolution of ageing and longevity. Proc R Soc Lond B 205:531–546

Kirkwood TBL, Rose MR (1991) Evolution of senescence: late survival sacrificed for reproduction. Phil Trans R Soc Lond B 332:15–24

Kowald A, Kirkwood TBL (1996) A network theory of ageing: the interactions of defective mitochondria, aberrant proteins, free radicals and scavengers in the ageing process. Mutat Res 316:208–236.

Lithgow GJ, White TM, Melov S, Johnson TE (1995) Thermotolerance and extended lifespan conferred by single-gene mutations and induced by thermal stress. Proc Natl Acad Sci USA 92:7540–7544

Luckinbill LS, Clare MJ (1985) Selection for life span in Drosophila melanogaster. Heredity 58:9–18

Masoro EJ (1992) Retardation of aging processes by nutritional means. Ann NY Acad Sci 673:29–35

Medawar PB (1952) An unsolved problem of biology. London, H K Lewis.

Orr WC, Sohal RS (1994) Extension of life-span by overexpression of superoxide dismutase and catalase in Drosophila melanogaster. Science 263:1128–1130

Partridge L, Barton NH (1993) Optimality, mutation and the evolution of ageing. Nature 362:305–311

Rose MR (1984) Laboratory evolution of postponed senescence in Drosophila melanogaster. Evolution 38:1004–1010

Schächter F, Cohen D, Kirkwood TBL (1993) Prospects for the genetics of human longevity. Human Genet 91:519–526

Service PM (1987) Physiological mechanisms of increased stress resistance in Drosophila melanogaster selected for postponed senescence. Physiol Zool 60:321–326

Smith DWE (1994) The tails of survival curves. BioEssays 16:907–911

Weindruch R, Walford RL (1988) The retardation of aging and disease by dietary restriction. Charles Thomas, Springfield

Williams GC (1957) Pleiotropy, natural selection and the evolution of senescence. Evolution 11:398–411

Zwaan BJ, Bijlsma R, Hoekstra RF (1995) Direct selection on lifespan in Drosophila melanogaster. Evolution 49:649–659

Emergence of Centenarians and Super-centenarians

B. Jeune and V. Kannisto[*]

Abstract

Until about 1950 centenarians were quite rare, and their number grew only slowly. Examinations of reported centenarians in the Nordic countries indicate that most reports of centenarians in the last century were not genuine. It is plausible that there were no true centenarians prior to 1800. However, since 1950 their number has more than doubled every 10 years in low-mortality countries like Western Europe and Japan. Women outnumber men at age 100 by four to one and, at still higher ages, even more. The main proportion of the rate of growth is due to a decrease in oldest-old mortality, which has continued at an annual rate of 1 to 2 %. Not only are there more centenarians today than ever, but they also live longer, approximately two years on average. As a result, more persons are now living to such high ages as 105 and 110. The proportion of centenarians reaching age 105 in the Nordic countries is increasing, especially since 1970. Semi- and super-centenarians may be as common among our grandchildren in the next century as centenarians are today in low-mortality countries.

Introduction

The demography of oldest old, including centenarians, was initiated by the French demographer Paul Vincent (1951) and his colleague Francoise Depoid (1973). This research has since been pursued by among others, the British demographer Roger Thatcher (1981, 1987) and the Finnish demographer Vaino Kannisto (1988, 1994).

The data on old age mortality in 30 countries collected by Kannisto and the data from England and Wales collected by Thatcher are now established as The Kannisto-Thatcher Database on Oldest Old at the University of Odense in Denmark. This database includes valid data on centenarians from 12 countries of Western Europe and Japan since 1950.

---

[*] Center for Research on Aging, Odense University Medical School, Winsløwparken 17, DK-5000 Odense C, Denmark

J.-M. Robine et al. (Eds.)
Longevity: To the Limits and Beyond
© Springer-Verlag Berlin Heidelberg New York 1997

Historical data on centenarians from the Nordic countries have now been collected, and are included in the The Danish Centenarian Register, which was established by Bernard Jeune and Aksel Skytthe at the University of Odense in 1995. This register reported data on centenarians back to the 1700s.

We present here some figures and tables from The Kannisto-Thatcher Database on Oldest Old and The Danish Centenarian Register. First we present figures on the proliferation and survival of centenarians in developed countries and then figures on the historical development of centenarians in the Nordic countries, especially in Denmark. Next we discuss the question of when it is plausible that the first centenarians, semi-super-centenarians and super-centenarians emerged in history.

Proliferation and Survival of Centenarians

Until about 1950 centenarians were quite rare and their numbers grew slowly. Around this time, however, their numbers in low-mortality countries began to increase considerably, with the main reason being a decline of mortality at ages over 80 years (Kannisto 1994; Vaupel and Jeune 1995).

In 12 countries of Western Europe for which we have reliable data at least since 1950, their number has more than doubled every 10 years (see Table 1). In Japan, where the number was small to begin with, it has grown relatively even faster.

We have processed these data using the method of extinct generations pioneered by the French demographer Paul Vincent (1951). This method is the most reliable way to calculate old age mortality if the age information on deaths is accurate. This proliferation of centenarians is likely to continue rapidly in the future, while the relative increase will, of course, be smaller because of higher base numbers.

Table 1. Centenarians in 12 Countries of Western Europe[a] and in Japan

Year	Persons aged 100 and over on January 1		Decennial growth factor	
	Western Europe	Japan	Western Europe	Japan
1950	733	72		
			2.2	2.1
1960	1584	154		
			2.1	2.2
1970	3254	345		
			2.1	2.7
1980	6918	918		
			2.2	3.4
1990	15301	3126		

[a] Austria, Denmark, England and Wales, Finland, France, Germany (FRG), Iceland, Italy, the Netherlands, Norway, Sweden and Switzerland.

Table 2. Proportion of Centenarians in Total Population, 1960 and 1990

Country	1/1/1960 Number	Per million	1/1/1990 Number	Per million
Austria	25	3.5	232	29.8
Belgium	...	...	474	48.1
Denmark	19	4.1	323	62.8
England and Wales	531	11.6	3890	76.3
Estonia	...	...	42	26.7
Finland	11	2.5	141	28.3
France	371	8.1	3853	67.9
West Germany	119	2.2	2528	40.0
Iceland	3	17.0	17	66.7
Ireland	...	...	87	24.8
Italy	265	5.4	2047	35.5
Japan	155	1.7	3126	25.3
Netherlands	62	5.4	818	54.7
New Zealand	18	7.6	198	59.2
Norway	73	20.4	300	70.7
Portugal	...	...	268	27.2
Singapore	...	...	41	15.2
Sweden	72	9.6	583	68.1
Switzerland	29	5.4	338	50.4
14 countries	1.753	5.3	18.394	45.1
19 countries	...	...	19.306	44.3

The proportion of centenarians in the total population, shown in Table 2, is still very small, less than 100 per million or one centenarian per 10 000 people. The ratio is highest in England, France and Scandinavia and lowest in countries where the population is still young, as in Japan and Singapore, or where mortality is relatively high, as in Estonia, Ireland and Portugal. In China, among the ethnic Chinese whose age data are accurate, the proportion is less than 5 per million but likely to grow rapidly (Zeng Yi, personal communication).

Figure 1 shows the centenarians as part of the oldest-old population. The black parts of the bars show the increase in numbers during the latest 20-year period. The largest increases have taken place at ages only a little over 80 years, but the relative growth is higher at older ages. The centenarians are only a small minority of the old people, but when we show them in larger scale (the small figure), we observe a virtually explosive growth.

Not only are there more centenarians today than ever, but they also live longer. The life expectancy of a 100-year-old woman increased in the same period from 1.8 to 2.1 years, and that of a man from 1.6 to 1.8. As a result, more persons are now living to such high ages as 105 and 110, as shown in Table 3. Even so, of 1000 centenarian women only two reach the age of 110, and of 1000 centenarian men fewer than 1.

How far is the present increase in life expectancy likely to advance? We do not know, of course, but if in the 13 countries of best quality data the mortality

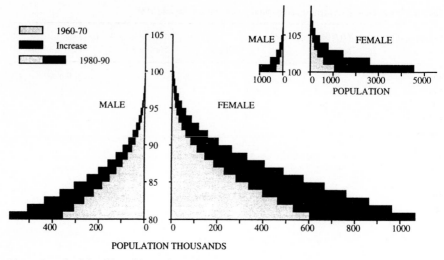

Fig. 1. Growth of the oldest old population from thirteen countries with good data.

decline of the last 20 years is repeated three more times or, in other words, if the decline continues at present speed until the year 2050, an 80-year-old woman can expect to live 12 more years (instead of less than 8 now), to age 92. A man could expect to live 9 more years, to 89. Because of mortality before age 80, the life expectancy at birth would be somewhat shorter. This is, however, strictly an extrapolation of observed age-specific mortality. The course of events may be different: the progress may accelerate or slow down, scientific breakthroughs may increase survival more than this, while unforseen developments may either increase or shorten it.

All our examples show that women live longer than men. This seems to be a universal rule when living conditions are not particularly harsh for women. In

Table 3. The Road from 100 to 110 Years. Thirteen countries with good data

Sex and period	Probability of surviving the age interval		
	100 to 105	105 to 110	100 to 110
Male			
1950–60	0.039269	–	–
1960–70	0.034054	0.014170	0.000482
1970–80	0.040671	0.006337	0.000258
1980–90	0.056520	0.014768	0.000835
Female			
1950–60	0.048129	–	–
1960–70	0.056046	0.011611	0.000650
1970–80	0.061910	0.020056	0.001242
1980–90	0.078088	0.027754	0.002167

Table 4. Sex Ratio of Centenarians. Thirteen countries with good data

Period and age	Number reaching age		Females per male
	Male	Female	
1980–1989			
100	13.066	55.761	4.1
101	7.048	32.349	4.6
102	3.732	18.150	4.9
103	1.958	10.035	5.1
104	963	5.295	5.5
105	482	2.723	5.6
106	236	.325	5.6
107	109	645	5.9
108	50	297	5.9
109	18	115	6.4
110+	7	57	8.1
Total	27.669	126.752	4.6
1970–1979	15.136	59.504	3.9
1960–1969	7.061	28.218	4.0
1950–1959	3.184	12.086	3.8

our group of 13 countries, women outnumber men (see Table 4) at age 100 by 4 to 1 and by even more at still higher ages.

This female advantage in survival has increased further in recent decades, but this should not be taken as an inevitable, natural result of mortality decline. Men's length of life has been shortened, e.g., by smoking, whereas most of the women who are now very old have never smoked. It has also been observed that women have been following advice on healthier living habits better than men, but these matters are subject to change.

Centenarians in the Past

The increase in the number of centenarians in low-mortality countries was very low in the first half of this century. This is evident from reliable data on centenarians from the Nordic countries of Denmark, Finland, Norway (only recently reliable) and Sweden (see Figure 2).

Figure 3 shows that the proportion of centenarians in Denmark increased from an estimated prevalence of one to two in one million in the last half of the 1800s to about 80 in one million in 1995. However, it grew very slowly in the first half of this century. It is only after 1950 that the number of living centenarians is higher than the annual number of centenarian deaths. This increase is much more pronounced among women than among men, leading to a female/male ratio of three to one.

It is, however, amazing that the number of reported centenarians decreased in the last century in Denmark. Table 5 shows that the number of reported cente-

Number

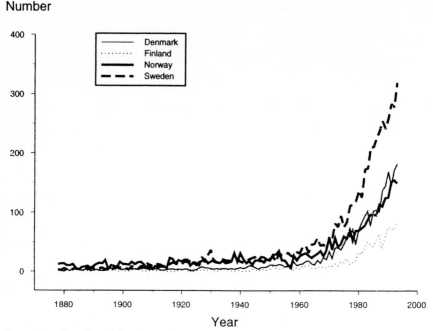

Fig. 2. Annual number of females reaching age 100 in the Nordic countries.

narian deaths was much higher in the first half than in the second half of the 1800s, especially among men, where the number fell from about 50 in a decade to less than 10 in a decade (the reported number in the period 1775–1804 did not cover all centenarian deaths, because some church books from this period were not available). This led to an increase in the female/male ratio.

By going through the church books from all the parishes in the County of Funen, which is the County in which Odense University is situated, we have registered all centenarian deaths in this part of Denmark (see Fig. 4). However, many church books from the period 1644–1814 were lost. This in fact partly explains the increase up to 1814. After this year it is evident that the number decreases.

Among the 274 reported deaths, 41 were reported as having died at the age of 105 years or older (see Table 6). The proportion of reported deaths at the age of 105 years or older among all reported centenarian deaths in the period 1644–1899 was almost four-fold higher (15 %) than the same proportion (3.9 %) in the period 1970–1993, and two-fold higher among men (20.7 %) than among women (10.8 %) in the early period, whereas the opposite was the case in modern times (2.3 versus 4.5 %).

Among the 41 cases who were reported to be 105 or older, eight were reported to be 110 or older (up to 130 years), whereas the first who reached the age of 110 in Denmark after 1900 did so in November 1994, and she is still the

Centenarians per million

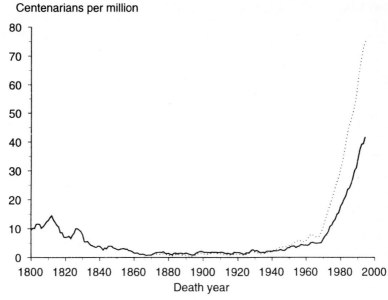

Fig. 3. Number of centenarians per million inhabitants in Denmark, 1800–1994 (five years moving average). Solid line, deaths. Dotted line, estimate of population of centenarians on the first of January.

only super-centenarian from this century in Denmark (Jeune et al. 1996). Only in two cases of the 41 could a birth registration with the same name be found in the year in which they supposedly were born if their age reported at death was true, and even this is not sufficient proof that they really reached these very high ages.

This examination of reported centenarians indicates that, before 1850, most reports of centenarians were not genuine, which also is in agreement with the fact that around 1850 the statistical bureau introduced different procedures for validation of reported ages at death.

A backward extrapolation of a regression line on the basis of the number of centenarians from the period 1870–1925 supports this conclusion. It shows a much lower number in the first half of the 1800s than reported, indicating that the first Danish centenarian emerged after 1780 (see Fig. 5).

The Swedish demographer Lundström has found the same decrease in the reported number of centenarian deaths in Sweden in the last century (Lundström 1995). However, the decrease began earlier in the 1800s, at a period where procedures for age-validation were introduced. Many other researchers have found signs of age exaggeration both in the past (Thoms 1873; Ernest 1938; Charbonneau and Desjardins 1990) and in the present (Mazess and Forman 1979; Palmore 1984). The classic work on this question is the investigation of reported centenarians about 1870 of Thoms (1973), who also contested the validity of extremely old people like Jenkins and Parr.

The first who investigated whether reports of centenarians before 1800 were genuine were the Canadian demographers Charbonneau and Desjardins (1990).

Table 5. Number of Reported Centenarian Deaths in Denmark.

Period	Females	Males	Total	Sex Ratio F:M
1775–1784	49	28	77	1.75
1785–1794	62	34	96	1.82
1795–1804	77	20	97	3.85
1805–1814	74	51	125	1.45
1815–1824	52	34	86	1.53
1825–1834	64	28	92	2.29
1835–1844	24	16	40	1.50
1845–1854	31	17	48	1.82
1855–1864	22	9	31	2.44
1865–1874	16	3	19	5.33
1875–1884	22	6	28	3.67
1885–1894	22	6	28	3.67
1895–1904	34	9	43	3.78
1905–1914	39	7	46	5.57
1915–1924	31	12	43	2.58
1925–1934	48	23	71	2.09
1935–1944	53	32	85	1.66
1945–1954	85	48	133	1.77
1955–1964	119	87	206	1.37
1965–1974	218	103	321	2.12
1975–1984	600	257	857	2.33
1985–1994	1.288	432	1.720	2.98
1775–1994	3.030	1.262	4.292	2.40

Among 178 reported Canadian centenarians from the province of Quebec, they found only one who could not be refuted. All of the others who could be verified died at an age lower than 100 at an average age of 88.

All these studies show that age exaggeration was more common among men than among women, leading to a sex ratio that was not expected, and to reporting of extremely high ages among men in periods where the number of centenarians was very small, given the small size of the population and the high mortality level, especially among old people. The number of reported supercentenarians in the 1700s and the 1800s and even today in certain regions is not believable, given the fact that the highest documented ages today are 121 years for women (Allard et al. 1994) and 113 for men (Wilmoth et al. 1996).

However, other studies have found evidence of centenarians before 1800. Kjærgaard (1995) in Norway, Hynes (1995) in England and Wales and Zhao (1995) in China found few reports of centenarians who might have been genuine. However, it is very difficult to reconstruct family and life history in the pre-industrial period because corroborative documents were very few.

Other, more statistical approaches than the historic-demographical approach can elucidate the question of the emergence of centenarians (Thatcher 1995; Vaupel and Jeune 1995; Wilmoth 1995; Zhao 1995). The comprehensive statistical

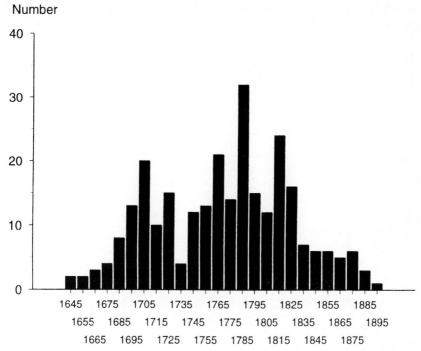

Fig. 4. Number of centenarian deaths found in the parish records on Funen from the period 1644–1899. Each bar represents counts of a decade.

analysis of Wilmoth (1995) shows that, using two criteria of emergence (at least one centenarian per century and at least one living centenarian on average at any time) and a variety of plausible mortality scenarios, "the best estimate indicates that the emergence of centenarians should have occurred by around 2500 B.C. in a world population of some 100 million people."

However, Wilmoth himself emphasizes that this conclusion is very sensitive to the assumption of the average survival beyond the age of 50 in the pre-industrial period. If the average survival was nearer 12 than 14 years throughout this period, as age determination of Medieval skeletons from Denmark indicates (Boldsen 1995), then it is plausible that there were no true centenarians prior to 1800, as postulated by Jeune (1994, 1995).

Table 6. Number of Reported Deaths at Age of 105+ Years. In parentheses, percentage of all centenarian deaths for the period 1644–1899 on Funen and the period 1970–1993 in Denmark

Period	Females		Males		Total	
	N	Pct	N	Pct	N	Pct
1944–1899	17	(10.8)	24	(20.7)	41	(15)
1970–1993	84	(4.5)	16	(2.3)	100	(3.9)

Number

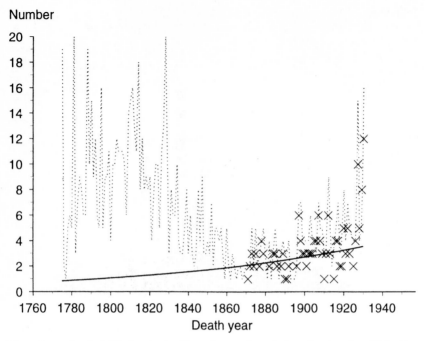

Fig. 5. Regression (solid line) on basis of the estimates of population of centenarians (X) for the period 1870–1925. The dotted line shows the number of centenarians deaths in the period 1775–1925.

Centenarians Reaching Age 105 or More

Whatever the truth is, the proliferation of centenarians is a very recent phenomenon in low-mortality countries (Kannisto 1994). Vaupel and Jeune (1995) have shown that the increase in births and the improved survival from birth to age 50 and from age 50 to age 80 only explained a minor proportion of the rate of growth (see Table 7). This is evident in the Nordic countries like Denmark, Norway and Sweden. The main proportion is due to a decrease in oldest-old mortality since 1950, which has continued at an annual rate of 1 to 2 %.

Figure 6 shows that even for birth cohorts from the first half of the last century the possibility of becoming a centenarian was very low: about one or two in 3000 among men and women. This possibility increases first for birth cohorts from 1860s or later, reaching about five in 1000 for women and two in 1000 among men, which is almost a 10-fold increase. For these birth cohorts even the proportion of centenarians reaching age 105 (semi-super-centenarians) in Denmark, Finland and Sweden is increasing, especially since 1970 (see Figure 7). (The reported number reaching age 105 in Norway has not been reliable until recently).

It could therefore be suggested that centenarians emerged after 1800 and proliferated after 1950, semi-super-centenarians emerged after 1900 and prolifer-

Table 7. Average Rate of Growth from the 1970s to the 1980s in the Number of Persons Attaining Age 100. The proportion of this rate of growth due to improved survival from age 80 to 100 and other factors

Country	Annual average rate of growth (in %)	Proportion (in %) of this rate of growth due to:				
		Improved survival from:			Decrease in net emigration	Increase in births from the 1870s to the 1880s
		Age 80 to age 100	Age 50 to age 80	Birth to age 50		
Austria	5.7	60	...	...	...	...
Belgium	7.8	80	...	...	...	...
Denmark	7.7	66	13	5	0	16
England and Wales	5.8	72	...	...	...	...
Finland	11.2	81	...	...	...	...
France	6.3	73	...	...	...	...
Japan	10.2	67	...	...	...	...
Norway	5.1	65	10	8	3	14
Sweden	7.1	66	12	10	9	4
Switzerland	9.2	73	...	...	...	...
West Germany	9.1	72	...	...	...	...

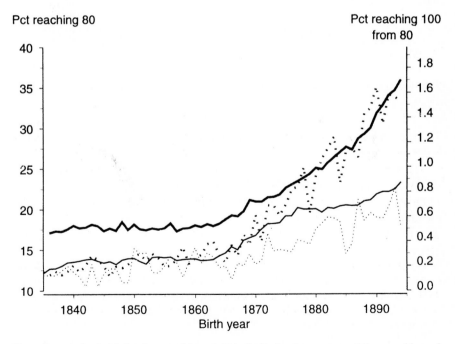

Fig. 6. Percentage of a birth cohort reaching age 80 (solid line) and percentage of 80-years-olds reaching age 100 (dotted lines). Thin lines are for males, thick lines for females.

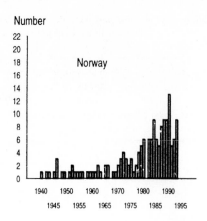

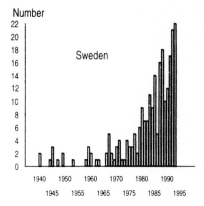

Fig. 7. The increase in the number of centenarians reaching 105 years in the Nordic countries.

ated after 1970, and super-centenarians emerged after 1950 and might proliferate in the next century.

Extremely few of our grandparents endured a century but centenarians may be commonplace among our grandchildren (Vaupel and Gowan 1986), and semi-super-centenarians may be as common among them as centenarians are today in low-mortality countries.

References

Allard M, Lèbre V, Robine JM (1994) Les 120 ans de Jeanne Calment. Le cherche midi éditeur. Paris
Boldsen JL (1995). Patterns of advanced age mortality in the medieval village tirup. In: Jeune B, Vaupel JW (eds) Exceptional longevity. From prehistory to the present. Odense Monographs on Population Aging, No. 2, Odense University Press, Odense
Charbonneau H, Desjardins B (1990) Vivre cent ans dans la vallée du Saint-Laurent avant 1800. Ann Démographie Historique 217–26

Depoid F (1973) La mortalité des grands vieillards. Population (Paris) 28:755–792

Ernest M (1938) The longer life. Adam & Co, London

Hynes J (1995) The oldest old in pre-industrial Britain: centenarians before 1800 – fact or fiction? In: Jeune B, Vaupel JW (eds) Exceptional longevity. From prehistory to the present; Odense Monographs on Population Aging, No. 2, Odense University Press, Odense

Jeune B (1994) Morbus centenarius or sanitas longaevorum? A critical review of centenarians studies. Population Studies of Aging nr. 15. Center for Health and Social Policy, Odense University, Odense

Jeune B (1995) In search for the first centenarians. In: Jeune B, Vaupel JW (eds) Exceptional longevity. From prehistory to the present. Odense Monographs on Population Aging, No. 2, Odense University Press, Odense

Jeune B, Skytthe A, Vaupel JW (1996) The demography of centenarians in Denmark (in Danish). Ugeskrift for Læger 158:7392–7396

Kannisto V (1988) On the survival of centenarians and the span of life. Population Studies 42:389–406

Kannisto V (1994) Development of old age mortality in 1950–1990. Evidence from 28 low-mortality countries. In: Jeune B, Vaupel JW (eds) Odense Monographs on Population Aging, No. 1, Odense University Press, Odense

Kjærgaard T (1995) Alleged Danish centenarians before 1800. In: Jeune B, Vaupel JW (eds) Exceptional longevity. From prehistory to the present. Odense Monographs on Population Aging, No. 2, Odense University Press, Odense

Lundström H (1995) Record longevity in Swedish cohorts born since 1700. In: Jeune B, Vaupel JW (eds) Exceptional longevity. From prehistory to the present. Odense Monographs on Population Aging, No. 2, Odense University Press, Odense

Mazess RB, Forman S (1979) Longevity and age exaggeration in Vilcamba. J Gerontol 34:94–98

Palmore EB (1984) Longevity in Abkhazia: a reevalution. The Gerontologist 24:95–96

Thatcher AR (1981) Centenarians. Populations Trends 25:11–14

Thatcher AR (1987) Mortality at the highest ages. J Inst Actuar 114:327–338

Thatcher AR (1995) A note on some historical data on old age mortality. In: Jeune B, Vaupel JW (eds) Exceptiional longevity. From prehistory to the present. Odense Monographs on Population Aging No. 2, Odense University Press, Odense

Thoms WJ (1873) Human longevity. Its facts and its fictions. John Murray, London

Vaupel JW, Gowan AE (1986) Passage to Methusalem: some demographic consequences of continued progress against mortality. Am J Public Health 76:430–433

Vaupel JW, Jeune B (1995) The emergence and proliferation of centenarians. In: Jeune B, Vaupel JW (eds) Exceptional longevity. From prehistory to the present. Odense Monographs on Population Aging, No. 2, Odense University Press, Odense

Vincent P (1951) La mortalité des vieillards. Population (Paris) 6:181–204

Wilmoth J (1995) The earliest centenarians: a statistical analysis. In: Jeune B, Vaupel JW (eds) longevity. From prehistory to the present. Odense Monographs on Population Aging No. 2, Odense University Press, Odense

Wilmoth J, Skytthe A, Friou D, Jeune B (1996) The oldest man ever? A case study of exceptional longevity. The Gerontologist 36:783–788

Zhao Z (1995) Record longevity in Chinese history – evidence from the Wang genealogy. In: Jeune B, Vaupel JW (eds) Exceptional longevity. From prehistory to the present. Odense Monographs on Population Aging, No. 2, Odense University Press, Odense 1995

A Demographic and Health Profile of Centenarians in China

Wang Zhenglian, Zeng Yi, B. Jeune, and J. W. Vaupel[*]

Introduction

It is more and more important for human beings to understand the demographic patterns and health status of centenarians because populations in most countries are aging. China's population, in particular, is aging at an extraordinarily rapid pace (Banister 1990; Ogawa 1988; Zeng and Vaupel 1989). Centenarians used to be exceedingly rare. They are still highly unusual, but the population of centenarians is doubling every decade or so. For example, from the 1970s to 1980s the average annual growth rates in the number of people attaining age 100 were 10.2, 9.2 and 9.1 % in Japan, Switzerland, and West Germany, respectively (Vaupel and Jeune 1995, p. 112). If current rates of mortality improvement persist, then it will be as likely for a child today to reach age 100 as it was for a child eight decades ago to reach age 80 (Vaupel and Gowan 1986).

China offers an unparalleled opportunity for studies of centenarians in a developing country for two key reasons. First, the Chinese population is so huge, now totalling more than 1.25 billion, that despite high mortality in the past there are large numbers of elderly Chinese. The 1990 census counted 6,681 centenarians,[1] 64,532 persons aged 95 or over, 416,134 persons aged 90 or over, and 2.32 million persons aged 85 or over. The annual growth rate of centenarians in China between 1982 and 1990 was 7.1 %. Second, age reporting in China appears to be generally reliable. In some minority subpopulations age exaggeration inflates counts of the very old. For the Han population of China, which accounts for 92 % of the total population, age reports are generally accurate, because Han Chinese, even if illiterate, can supply a precise date of birth (Coale and Li 1991, pp. 294). This is important because misreports of age distort demographic analyses of the very old in most developing countries as well as in the United States and some other developed countries.

[*] Wang Zhenglian is a Ph. D candidate of Odense University and Data base manager and researcher of Max Planck Institute of Demographic Research (MPIDR). Zeng Yi is Professor of Peking University and Distinguished Research Scholar of MPIDR. Bernard Jeune and James W. Vaupel are Professors at Medical School of Odense University. James W. Vaupel is also director of MPIDR. All correspondence should be addressed to: Wang Zhenglian, MPIDR, Doberaner Str. 114, D-18057, Rostock, Germany.
[1] Twenty-four percent of the 6681 reported centenarians are minority nationalities who may have overstated their ages, as will be discussed later.

J.-M. Robine et al. (Eds.)
Longevity: To the Limits and Beyond
© Springer-Verlag Berlin Heidelberg New York 1997

To our knowledge, there have been very few centenarian studies, limited to certain local areas, and no detailed studies of centenarians on a national scale in China so far. Based on the most recent Chinese census data and the centenarian surveys conducted in Hangzhou and Beijing by Wang Zhenglian, this paper intends to provide a profile of demographic characteristics and self-reported health status of Chinese centenarians, including data quality, age, sex and regional distributions, education, occupation, living arrangements, smoking and alcohol consumption, self-reported health status, etc.

Data Quality of Chinese Centenarians

Coale and Li (1991) analyzed the data quality at advanced ages from the 1982 Chinese census, with the Han and the ethnic minority populations combined and age 100 and above combined, because the detailed tabulations by single years of age of Han and minority nationalities were not available when they conducted the study. Nevertheless, by studying Xinjiang province, where Weiwuer and other ethnic groups consist of about 60 % of the province, they discovered that the elderly of minority populations seriously overstated their ages. In 1982, the reported numbers of centenarians in Xinjiang consisted of 22.5 % of the total number of centenarians in China as a whole, whereas the total population of Xinjiang consisted of only 1.3 % of the total population of China. As discovered by Coale and Li (1991), there were 144 males listed as over 110 years of age in China in 1982; 121 were in Xinjiang and another 15 were in four other provinces with the highest proportion of minorities whose cultures were not related closely to the Han Chinese. The five provinces with the highest fraction of these minorities (Xinjiang, Guanxi, Qinghai, Ningxia, and Yunnan) contained less than 9 % of the population of China but 94.4 % of males listed as being over 110. The centenarian data and the death rates in 1982 were seriously distorted if the data for all China including Xinjiang were used, but they escaped such distortion if the data from Xinjiang were omitted (Coale and Li 1991, pp. 298–300).

Fortunately we have the 1990 census data set on centenarians with detailed information of single year of age and ethnic grouping. Again, the reported minority Chinese centenarians account for about 24 % of the total number of centenarians in China, but the minority population consists of only slightly less than 8 % of the total population. For the males listed over age 110, 85.4 % belong to the minority groups. As was observed by Coale and Li (1991), the demographic indexes for measuring age misreporting also clearly show that the reports of minority Chinese centenarians are mostly not true, because they seriously overstated their age.

On the other hand, Han Chinese, whatever their education level, know their birth dates precisely. Young and educated people can supply their birth dates according to the Western calendar; the old and illiterate can supply their birth dates according to the traditional Chinese calendar. The Chinese calendar consists of a simple version of a cycle of 12 animal years, such as chicken, rat, tiger, etc., as well as a more complicated version of a cycle of 60 years, with 12 animals and each animals five different qualities. According to the Han Chinese traditions,[2] the precise date of birth is significant in making decisions on important life events such as marriage matching, date of marriage, date to start to build a house, and date of travelling, etc. It is widely recognized that age reporting of Han Chinese elderly is highly accurate in general (see, for example, Coale and Li 1991; Vaupel et al., in preparation), but we did not have enough knowledge about the accuracy of the data of Han Chinese centenarians before the detailed data became available for this study. Therefore, an evaluation of data quality of Han Chinese centenarians follows.

The population sizes and sex ratios of Han Chinese elderly aged 80 and above appear reasonable, as compared with those of Japan and Sweden (see Table 1). Table 2 presents an index for measuring the accuracy of age reporting between ages 65 and 105. The index was proposed by Coale and Li (1991), and it is computed as a mean of the ratios of the number at each age to a two-stage moving average (the five-term average of a five-term average) from age 65 to 105. Sweden is considered to be the country with the best data accuracy in the world so, the more accurate the age reporting, the closer the mean indexes are to the Swedish ones. Table 1 shows a very close match of the Han Chinese indexes to the Swedish indexes.

The above analysis demonstrated that there is no serious age heaping and that age reporting is of high quality in general in the Han Chinese elderly population, although it is still not as good as in Sweden. However, the absence of significant digit preference at ages divisible by ten or five is not necessarily proof of data accuracy, since other kinds of errors of age reporting may also distort the data quality. Do any kinds of systematic age misreporting other than heaping, such as exaggeration of age for seniority honour, exist in the Han Chinese elderly population? One way of addressing this question is to imagine that, if age exaggeration at very old ages is serious, the reported total number of very old persons must be relatively large as compared with the reported total number including younger persons, and the ratio of life expectancy at very old age to the conditional survival probability at middle age should be relatively large. As shown by Coale and Kisker (1986, p. 398), the ratios of those aged 95 years or over to those aged 70 or over in the 23 countries with accurate data quality were all less than six per thousand, whereas the ratios for the 28 countries with poor data clearly showed the exaggeration of very old persons aged 95 or over, extending from 1 % to 10 %. This ratio for male and female Han Chinese in 1990 is 0.76 per thousand and 2.18 per thousand, respectively, which is almost exactly the same as their Swedish counterparts in the period from 1985–1994. Coale and Kisker calculated

---

[2] It is important to ask for the date of birth and to compute age by subtracting from the date of the census or survey (if respondents supply the Chinese calendar, conversion to the Western calendar is needed). If the questionnaire asks the individual's age, the Chinese system of reckoning nominal age makes the response ambiguous, because a person may be counted as one-year-old on the day of birth and one year older with each new year according to the Chinese tradition.

Table 1. Male and female population, sex ratio (SR) at advanced ages for Han Chinese (1990), Japanese (1985–1990) and Swedes (1985–1994)

Age	Han Chinese			Japanese			Swedes		
	Male	Female	SR	Male	Female	SR	Male	Female	SR
80– 84	1845877	3155219	58.50	3109384	5067146	61.36	759811	1237222	61.41
85– 89	555217	1227726	45.22	1297711	2520394	51.49	329311	676627	48.67
90– 94	82920	240949	34.41	323315	765460	42.24	93760	241729	38.79
95– 99	10303	28606	26.69	44.289	128484	34.47	15495	49493	31.31
100–104	779	3611	21.56	2803	11106	25.24	1291	5109	25.27
105–109	94	447	21.11	86	466	18.45	48	246	19.51

the ratio of T_{100}/T_{70} (ratio of total person-years above age 100 to the total person-years above age 70) for various countries relative to T_{100}/T_{70} for Sweden in 1980. The ratios in the countries with accurate data were mostly below 1, with the lowest value being 0.04 for Finland in 1950, and a few cases exceeding 1 but not more than 1.28. The ratios in the countries with poor data ranged from 7.92 (Mauritius) to 82.1 (Dominican Republic), which shows clearly the overstatement of age over 100 years in these countries. The ratio of T_{100}/T_{70} for Han Chinese in 1990 relative to T_{100}/T_{70} for Sweden in 1980 is 0.13, which demonstrates again that the Han Chinese are among the populations with accurate data quality for old people.

Although we trust that the age reporting of Han centenarians is accurate in general, we also found that the super-centenarians' age reporting is questionable. The ratio of persons aged 105 or over to persons aged 100 to 104 is 0.14 for Han Chinese, but 0.04–0.05 for 13 European countries and Japan combined. Is this due to age exaggeration at age 105 and over or are death rates of Han Chinese super-centenarians lower because of much higher selection at ages 105 and over? We think that both factors may play a role. Note that the population size of super-centenarians aged 105 and above is very small, and age overstatement even by a small number of people can seriously distort the data quality of this age group. The super-centenarians account for only a small portion of the total centenarian population, so that its questionable number may not negate our conclusion that the data quality of Han Chinese centenarians is generally good.

The field observations in the centenarian surveys in Hangzhou and Beijing by Wang Zhenglian[3] also confirm that the age reporting of Chinese centenarians

---

[3] As part of her Ph. D research, Wang Zhenglian just completed the field work of the centenarian surveys in Hangzhou and Beijing municipalities in China. Hangzhou municipality has a total population of 5.92 million and is located in the southeastern part of the country. Beijing is the capital city of China with a total population of 11.4 million and is located in the northern part of China. Wang Zhenglian has personally visited 83 (40 in Hangzhou and 43 in Beijing) centenarians and done health examinations with the help of one medical doctor. As the second stage of her field survey work, Wang will conduct centenarian surveys in Chendu municipality and its surrounding rural areas, located in the southwestern part of China. The Chendu data are not included in this paper since the surveys were not completed when this paper was written.

Table 2. A comparison of the mean ratios of the number at each age to a two-stage moving average (the five-term average of a five-term average) from age 65 to 105 in China (1990) and Sweden (1985–1994)

	Male			Female			Both sexes		
	China	Sweden	Difference (%)	China	Sweden	Difference (%)	China	Sweden	Difference (%)
Survivors	0.853	0.884	−3.5	0.884	0.908	−2.7	0.875	0.902	−3.0
Deaths	0.891	0.916	−2.7	0.925	0.937	−1.2	0.916	0.934	−1.8

is generally good. During Wang's field surveys, she went through the following steps to validate the centenarians' age: 1) looking up the centenarians' household registration records and official certificate of elderly; 2) visiting the centenarians' neighbours; 3) checking with the ageing committee;[4] 4) asking each centenarian to personally provide his/her Chinese animal year at birth. All centenarians interviewed (all Han Chinese) could remember his/her animal year at birth clearly; 5) asking about some important historical events that occurred at the beginning of this century; and 6) asking about some demographic events that occurred in the centenarians' life, such as age of first marriage, ages at first and last child's birth, ages of centenarians' surviving and dead children, and age of the centenarians' surviving spouse or year in which his or her spouse died. Wang found one lady who reported her birth year as 1897, but it was misprinted as 1881 in her household registration. Wang found only one among the 83 Han Chinese interviewees who exaggerated her age by 10 years, and her correct age is probably 98 instead of the reported 108. According to Wang's filed observation, the main reason she overstated her age was that she wanted to show that her and her dead husband's Tai Ji Chinese exercise was good for longevity.

Since the Han Chinese centenarians' age reporting is generally reliable but the reported ethnic minority persons aged 100 or above in China are most likely not true centenarians, we will exclude the minority groups in the subsequent analyses in this paper.

Analysis Based on the 1990 Census Data

The following analyses are based on the 1990 census data for Han centenarians in China. Since the detailed tabulations and the raw data set for Han Chinese centenarians from the 1982 census are not available, we are not able to compare the demographic profiles of the centenarians in 1982.

---

[4] Ageing committees for helping elderly people and related administrative work were established in the middle 1980s in China at the national, provincial, municipal, district, county, township, and street neighbourhood committee levels.

Age and Sex Distribution

Table 3 presents the single year of age distribution and sex ratios of Han Chinese centenarians. Nearly 60 % of Han Chinese centenarians are just 100 or 101 years old. The percent shares reduce quickly after age 101, and persons of age 105 and over comprise 12.5 % of the total number of centenarians. The sharp decline of percent share after age 101 and the small proportion over age 105 are due to high mortality rates at extremely high ages. There is about one male Han centenarian per five female Han centenarians or 21 %. The sex ratio at ages 105–109 is 21.1 for Han Chinese, and it is 18.5 and 19.5 for Japan and Sweden, respectively. The sex ratio at age 110 and over is substantially higher than at the other ages (38.2 %), which may be due to random fluctuation resulting from the very small number or because more males overstated their age as being over 110. As discussed before, we think that the Han Chinese data quality at ages 105 and over is questionable. As in some other European countries with generally accurate population data, the Han Chinese data for extremely high ages, such as 105 and over, must be used with great caution.

Education

Table 4 presents the education distribution of Han Chinese centenarians. About 71 % of male centenarians and 93 % of female centenarians in the Han Chinese population in 1990 had no education. The percentage having a primary school education was 23 % for males and 2 % for females. There were 3 % and 2 % of male Han Chinese centenarians who had middle and high school educations, respectively. Among a total of 3977 female Han centenarians, only 13 had a mid-

Table 3. Age distribution of Han Chinese centenarians

Age	Male Number	Percent	Female Number	Percent	Both sexes Number	Percent	Sex ratio (100 × M/F)
100	311	36.85	1464	36.81	1775	36.82	21.2
101	189	22.39	874	21.98	1063	22.05	21.6
102	115	13.63	571	14.36	686	14.23	20.1
103	63	7.46	338	8.50	401	8.32	18.6
104	57	6.75	238	5.98	295	6.12	23.9
105	31	3.67	163	4.10	194	4.02	19.0
106	23	2.73	105	2.64	128	2.66	21.9
107	15	1.78	74	1.86	89	1.85	20.3
108	11	1.30	52	1.31	63	1.31	21.2
109	9	1.07	37	0.93	46	0.95	24.3
110	7	0.83	27	0.68	34	0.71	25.9
>110	13	1.54	34	0.85	47	0.97	38.2
Total	844	100.00	3977	100.00	4821	100.00	21.2

Table 4. Education of Han Chinese centenarians in China (1990)

	Male		Female		Total	
	#	%	#	%	#	%
No education	596	70.62	3867	97.23	4463	92.57
Primary school	195	23.10	97	2.44	292	6.06
Middle school	29	3.44	7	0.18	36	0.75
High school	18	2.13	2	0.05	20	0.41
University or higher	6	0.71	4	0.11	10	0.20
Total	844	100	3977	100	4821	100

dle school or higher education level. The very low education level among the Chinese centenarians, especially among the females, is mainly due to the very rare educational facilities and opportunities available about 90 years ago, when today's centenarians were children. The even much lower education level among the female centenarians as compared to the males reveals the low status of women in the old China, when the son preference was so strong that extremely few girls had an opportunity to go to school. It should be noted that we are not able to study the relationship between education level and longevity since we do not have education information for those who were born in the same years as the surviving centenarians but who died before 1990, when the census was conducted.

Regional Distribution

Table 5 present the regional distribution of centenarians among the Han Chinese population in China. It is interesting to note that the density of centenarians among Han Chinese is higher in the southern parts of China such as Guangxi, Guangdong, Hainan and Sichuan.[5] The less developed, northwestern parts of China, including Inner Mongolia, Xinjiang, Gansu, Qinghai, Ningxia, Shanxi, Shaanxi, and Tibet, have the lowest density of Han Chinese centenarians. Inner Mongolia, Xinjiang, Gansu, Qinghai, Ningxia, and Tibet are also the regions where the ethnic minority populations are concentrated, but we do not know the density of the minority centenarians since their age reporting is not reliable. The province with the highest density of Han centenarians is Guangxi, but it is not a region with advanced socio-economic development. Guangdong, which is one of the most economically developed provinces, has the second highest centenarian density. The three mostly urbanized municipalities of Beijing, Shanghai and

---

[5] The percentages of super centenarians aged 105 or over among total persons aged 100 or over are 12.9, 1.8, 9.5 and 13.8 in Guangxi, Guangdong, Hainan, and Sichuan, respectively. The sex ratios at ages 100 to 104 are 17.1, 10.4, 21.3, and 21.1 in Guangxi, Guangdong, Hainan, and Sichuan, respectively. The sex ratios at ages 105 or above are 21.2, 9.8, 0, and 33.8 in Guangxi, Guangdong, Hainan, and Sichuan, respectively. These data do not indicate a serious age overstatement in these four provinces with the highest density of centenarians.

Table 5. Regional distribution of centenarians (persons/million) among the Han Chinese population in China

Province	Male	Female	Total	Sex ratio (100 male/female)	Persons/million
Beijing	3	37	40	8.1	3.82
Tianjing	4	22	26	18.2	3.02
Hebei	13	81	94	16.0	1.56
Shanxi	9	38	47	23.7	1.64
Inner Mongolia	9	7	16	128.6	0.88
Liaoning	41	89	130	46.1	3.59
Jining	51	49	100	104.1	4.42
Heilongjiang	24	52	76	46.2	2.27
Shanghai	7	71	78	9.9	5.87
Jiangsu	41	223	264	18.4	3.94
Zhejiang	40	128	168	31.3	4.07
Anhui	50	258	308	19.4	5.51
Fujian	16	117	133	13.7	4.47
Jiangxi	23	88	111	26.1	2.94
Shandong	46	216	262	21.3	3.12
Henan	55	305	360	18.0	4.26
Hubei	40	133	173	30.1	3.33
Hunan	28	131	159	21.4	2.73
Guangdong	65	628	693	10.4	11.49
Guangxi	73	414	487	17.6	18.74
Hainan	10	53	63	18.9	10.44
Sichuan	137	602	739	22.8	7.16
Guizhou	21	86	107	24.4	4.47
Yunnan	14	64	78	21.9	3.09
Tibet	0	0	0		0.00
Shaanxi	17	53	70	32.1	2.14
Gansu	4	18	22	22.2	1.07
Qinghai	0	3	3	0	1.12
Ningxia	1	5	6	20.0	1.91
Xinjiang	2	6	8	33.3	1.31
China (whole)	844	3977	4821	21.2	4.64

Tianjing have the highest proportion of elderly aged 65 or above, but their densities of centenarians are not among the highest. No clear association between the density of centenarians and socio-economic development level is evident in China. What the socio-economic and environmental factors affecting people's longevity are remains an open question that deserves much more research.

The sex ratios of Han centenarians in all provinces seem reasonable except in Inner Mongolia, Jining, Liaoning, and Heilongjiang, the areas along the north boundary with the coldest weather in China. The high sex ratios in these four northern provinces, especially in Inner Mongolia and Jining, are probably due to some males who are less than 100 years old overstating their ages. The centenarian data from these provinces should be used with great caution, and further study of the data quality in these areas is needed.

Analysis Based on the Hangzhou and Beijing Centenarian Surveys

This section is based on data collected in two regional surveys of centenarians in Hangzhou and Beijing municipalities. Unlike the data presented in the previous section, which used the census data for China as a whole, the data and discussion in this section can only be interpreted as two case studies because they cannot represent the country as a whole.[6]

Occupation

Table 6 provides the occupation distribution of centenarians before age 65. For the male centenarians interviewed, the majority were engaged in non-agricultural work; only 17.6 % were farmers. However, only 14.6 % of the female centenarians were industrial, commercial or technical workers; 57.6 % were housewives plus 28.8 % worked in the farming fields. This is further evidence showing the low social and economic status of women in the old China.

Marital Status and Living Arrangements

Among the 83 interviewed centenarians in Hangzhou and Beijing, there are no currently divorced persons, there was one man and one woman who were never married, there are only four currently married men, and the rest are all widowed.

Table 6. Occupations of the surveyed centenarians in Hangzhou and Beijing before age 65

| | Hangzhou | | | | Beijing | | | | Total | | | |
| | Male | | Female | | Male | | Female | | Male | | Female | |
	#	%	#	%	#	%	#	%	#	%	#	%
Agricultural work	3	66.7	11	32.4	0	0	8	25.0	3	17.6	19	28.8
Industry worker	3	33.3	3	8.8	1	9.1	1	3.1	4	23.5	4	6.1
Commercial	0	0	2	5.9	2	18.2	0	0	2	11.8	2	3.0
Technical staff	0	0	0	0	6	54.5	3	9.4	6	35.3	3	4.5
Government officer	0	0	0	0	2	18.2	0	0	2	11.8	0	0
Housewife			18	52.9			20	62.5			38	57.6
Total	6	100	34	100	11	100	32	100	17	100	66	100

---

[6] Another caution that should be kept in mind is that, in some categories, the sample size is too small (e.g., only one or two observations) and no statistically significant tests are presented. This is because the main purpose of this paper is to provide a general picture of the demographic and health characteristics of the surveyed centenarians; the detailed statistical analyses will be conducted later.

Table 7. Living arrangements of the surveyed centenarians at the time of interview, in Hangzhou and Beijing

	Hangzhou				Beijing				Total			
	Male		Female		Male		Female		Male		Female	
	#	%	#	%	#	%	#	%	#	%	#	%
Spouse	0	0	0	0	4	36.4	0	0	4	23.5	0	0
Children	5	83.3	32	94.1	6	54.5	29	90.6	11	64.7	61	92.4
Other relatives	0	0	1	2.9	0	0	1	3.1	0	0	2	3.0
Nursing home	1	16.7	0	0	0	0	1	3.1	1	5.9	1	1.5
Alone	0	0	1	2.9	1	9.1	1	3.1	1	5.9	2	3.0
Total	6	100	34	100	11	100	32	100	17	100	66	100

Table 7 presents the living arrangements of the centenarians in Hangzhou and Beijing. The majority of centenarians in Hangzhou and Beijing live with children and/or grandchildren, especially the female centenarians. Only two centenarians live in an elderly nursing home, and three centenarians (3.6 % of the total) live alone. Three of the four male centenarians living only with their wives are either high-ranking officers or famous specialists who have a higher economic status and better housing conditions which may enable them to not rely on children's support. Most of the other three centenarians living alone or with spouse only have no children at all. It is clear that a large majority of the centenarians in Hangzhou and Beijing live with their children, which is due to the Chinese tradition requiring children to pay respect to and provide care for old parents, and due to the fact that elderly nursing home facilities are not yet commonly available.

Smoking and Alcohol

Half of the male centenarians and 12 % of the female centenarians in Hangzhou are smokers, but the fractions of smokers were 0 and 6 % among male and female centenarians, respectively, in Beijing (see Table 8). Slightly more than a third of the centenarians in Hangzhou drink alcohol, but this number was less than 5 % in Beijing (see Table 9). The reason there are many more smokers and alcohol drinkers among centenarians in Hangzhou than in Beijing is that 72 % of the centenarians in Hangzhou live in rural areas versus 16 % of their counterparts in Beijing, and smoking and alcohol drinking are more popular in rural areas than in urban areas. Another factor that may be useful in interpreting why there are many more alcohol drinkers in Hangzhou is that a famous Chinese Yellow wine (not strong) is produced and is popular in Hangzhou and its surrounding areas.

Table 8. Number and distribution of those smoking at the time of interview in Hangzhou and Beijing centenarian populations

	Hangzhou Male		Female		Beijing Male		Female		Total Male		Female	
	#	%	#	%	#	%	#	%	#	%	#	%
Smoking	3	50.0	4	11.8	0	0	2	6.3	3	17.6	6	9.1
Not smoking	3	50.0	29	85.3	11	100	30	93.8	14	82.4	59	89.4
No information			1	2.9							1	1.5
Total	6	100	34	100	11	100	32	100	17	100	66	100

Table 9. Number and distribution of those drinking alcohol at the time of interview in Hangzhou and Beijing centenarian population

	Hangzhou Male		Female		Beijing Male		Female		Total Male		Female	
	#	%	#	%	#	%	#	%	#	%	#	%
Drinking alcohol	2	33.3	12	35.3	1	9.1	1	3.1	3	17.6	13	19.7
Not drinking alcohol	4	66.7	21	61.8	10	90.9	31	96.9	14	82.4	52	78.8
No information			1	2.9							1	1.5
Total	6	100	34	100	11	100	32	100	17	100	66	100

Reported Health Status

The majority (about 80 %) of the centenarians in Hangzhou reported that they never had a serious disease in their lives. By contrast, the majority (about 80 %) of centenarians in Beijing reported that they have suffered serious disease (see Table 10). Indeed, one may draw a conclusion that the health of centenarians in Hangzhou is much better than of those in Beijing if one only looks at Table 10. However, the percentages of those bedridden in Hangzhou and Beijing are almost the same (see Table 11). The differences in percents of those able to count numbers and do simple computation, to draw a simple picture, to use chopsticks, or to pick up a coin from the floor are not large between Hangzhou and Beijing. How to interpret the contradictory information revealed in Tables 10, 11 and 12? Our explanation is that the real health status of centenarians in Hangzhou and Beijing may not differ too much, but the perception of a "serious disease" may differ substantially between these two regions. In Hangzhou, 72 % of centenarians live in rural areas and do not have access to modern medical facilities to check out a disease they may have. Some rural centenarians even reported that they had never had their blood pressure measured and did not know what high blood pressure disease was. It is natural for those rural centenarians to report that they had no serious disease if they did not know the disease existed.

Table 11 shows that more than two thirds of the centenarians did not need help for mobility. About one third of the centenarians could count numbers, do

Table 10. Number and distribution of those who ever suffered from serious disease (SD) in Hangzhou and Beijing centenarian populations

| | Hangzhou | | | | Beijing | | | | Total | | | |
| | Male | | Female | | Male | | Female | | Male | | Female | |
	#	%	#	%	#	%	#	%	#	%	#	%
Ever suffered SD	1	16.7	8	23.5	9	81.8	25	78.1	10	58.8	33	50.0
Never had SD	5	83.3	25	73.5	2	18.2	7	21.9	7	41.2	32	48.5
No information			1	2.9							1	1.5
Total	6	100	34	100	11	100	32	100	17	100	66	100

Table 11. Percentage distribution of centenarians needing help for mobility or if bedridden at the time of interview in Hangzhou and Beijing

| | Hangzhou | | Beijing | | Total | |
| | Male | Female | Male | Female | Male | Female |
	#	%	#	%	#	%
Bedridden	5	12.5	6	14.0	11	13.2
Mobility needs help	4	10.0	12	27.9	16	19.3
Mobility needs no help	31	77.5	25	58.1	56	67.5
Total	40	100	43	100	83	100

a simple computation, and pick up a coin from the floor. The data shown in Tables 11 and 12, plus the rich and concrete first-hand information obtained in the one-on-one personal interviews by Wang Zhenglian, confirm that the health status of centenarians in Hangzhou and Beijing is generally good.

About one third of Danish centenarians do not need help for mobility and 27.5 % of them can pick up a coin from the floor, according to an ongoing study supervised by Bernard Jeune, while the respective figures for the Han Chinese counterparts are 57.5 % and 35.6 %. Is the health status of Han Chinese centenarians better than that of the Danish? Our answer is, not necessarily, since the above figures are not based on sophisticated medical examination and they may be influenced by other socio-economic or cultural factors. For example, the much better facilities for assisting very old people in Denmark may lead more centenarians to rely on help for mobility, whereas most Chinese centenarians have to try their best to manage for themselves because no facilities are available. What the differences are in health status between Chinese and Western centenarians and what factors caused these differences are still open questions and much more research is needed.

Table 12. Percentage of centenarians able to perform some activities, in Hangzhou and Beijing

	Hangzhou Male #	Female %	Beijing Male #	Female %	Total Male #	Female %
Count numbers/computing	13	32.5	13	30.2	26	31.3
Draw a simple picture	4	10.0	9	20.9	13	15.7
Use chopsticks	40	100.0	30	70.0	70	84.3
Pick up a coin from the floor	12	30.0	12	27.9	24	35.6

Conclusions

Analyses presented in this paper confirm that the data quality of Han Chinese centenarians is generally good, but the Chinese ethnic minority populations seriously overstate their ages. Although generally accurate age reporting is evident among the Han Chinese centenarians, the data at extremely high ages, such as over age 105, must be used with great caution, as in some other developed countries with good data. The Han centenarian census data in Inner Mongolia and Jining appear not to be reasonable and to deserve further investigation.

There is about one male Han centenarian per five female Han centenarians, which reveals the significantly longer life span for females than males. That a large majority (71 % of the males and 93 % of the females) of Han centenarians are illiterate is mainly due to the lack of educational facilities nine decades ago. Among the female centenarians, 57.6 % were housewives plus 28.8 % worked in the farming fields, which again shows the very low social and economic status of women in the old China. The density of centenarians among Han Chinese is higher in the southern parts of China, such as Guangxi, Guangdong, Hainan and Sichuan. The less developed northwestern parts of China, including Inner Mongolia, Xinjiang, Gansu, Qinghai, Ningxia, Shanxi, Shaanxi, and Tibet, have the lowest density of Han Chinese centenarians. There is no clear evidence to show an established association between the density of centenarians and socio-economic development level, and this area deserves much more research.

Data collected in the Hangzhou and Beijing centenarian surveys show clearly that a large majority of the centenarians in Hangzhou and Beijing live with their children, which is due to the Chinese tradition requiring children to pay respect to and provide care for old parents, and due to the fact that elderly nursing home facilities are not yet commonly available. More than two thirds of the centenarians do not need help for mobility. About one third of the centenarians can count numbers, do a simple computation, and pick up a coin from the floor. The Hangzhou and Beijing centenarian surveys show that the health status of centenarians in Hangzhou and Beijing is generally good, but the differences in health status between Han Chinese and Western centenarians and their causal factors are still open questions and need much more research.

References

Banister J (1990) Implications of aging of China's population. In: Zeng Yi, Zhang Chunyuan, Peng Shongjian (eds) Changing family structure and population aging in China: a comparative approach. Beijing, Peking University Press, published in English. Also in: Dedley L Poston, Jr and David Yaukey (eds) The population of modern China. New York, Plenum Press

Coale A, Kisker EE (1986) Mortality crossovers: reality or bad data? Population Studies 40:389–401

Coale A, Shaomin Li (1991) The effect of age misreporting in China on the calculation of mortality rates at very high ages. Demography 28 (2):293–301

Ogawa N (1988) Aging in China: demographic alternatives. Asian-Pacific Population J 3 (3):21–64

Vaupel JW, Gowan AE (1986) Passage to Methuselah: some demographic consequences of continued progress against mortality. Am J Public Health 76:430

Vaupel JW, Jeune B (1995) The emergence and proliferation of centenarians. In: Jeune B, Vaupel JW (eds) Exceptional longevity: from prehistory to the present. Odense Monographs on Population Aging 2, Odense, Odense University Press

Zeng Yi, Vaupel JW (1989) Impact of urbanization and delayed childbearing on population growth and aging in China. Population Dev Rev 15 (3):425–445

Acknowledgements

The authors are very grateful to the State Statistical Bureau of China for providing the raw data set. Support from the National Center for Ageing Research of China, National Natural Science Foundation of China and Aging Research Unit at Medical School of Odense University is highly appreciated. We are very grateful to Professor Xiao Zhengyu at the National Center of Ageing Research and the Ageing Committees in Hangzhou and Beijing for their effective help.

Centenarians: Health and Frailty

*B. Forette**

The number of centenarians is rapidly growing. In France, their estimated number was 200 in 1953 and over 3000 in 1989 when the IPSEN study was initiated. They are expected to be 6000 at the turn of the century. In Britain, where the Queen sends congratulations to those of her subjects who enter their second century, Dr. Cyril Clarke wrote to the royal secretary for details. The reply stated that 300 messages had been sent in 1955, 1200 in 1970 and 3300 in 1987 (Clarke et al. 1994). The figures are more or less similar in other developed countries. To many gerontologists this evolution does not mean that the maximum life span of the human species has been extended, since centenarians existed in past centuries. Their larger present number might be the result of better living and health care conditions, which allow more people to reach the limits of the human genetic program. Other investigators observe that at least some aspects of the aging process seem to become slower in the newer cohorts (Svanborg et al. 1984).

Considering the magnitude of such a demographic phenomenon, studies on centenarians are still relatively scarce. Few studies include more than 100 subjects (Karasawa 1979; Allard 1991; Beregi 1990; Beard 1991; Poon 1992; Louhija 1994) and most of them are multidisciplinary, although they lay emphasis on some special field. Some studies are limited to particular aspects, such as dementia (Hauw et al. 1993; Powell 1994; Sobel et al. 1995), neuropathology (Hauw et al. 1986; Delaere et al. 1993; Bancher and Jellinger 1994; Fayet et al. 1994; Giannakopulos et al. 1995), thyroid function (Mariotti et al. 1993), body fat and metabolism (Paolisso et al. 1995; Barbagallo et al. 1995), immunology (Effros et al. 1994; Franceschi et al. 1995), blood clotting (Mari et al. 1995), special treatments (Cobler et al. 1989; McCann and Smith 1990), or autopsies (Klatt and Meyer 1987).

When investigating centenarians, one of the main difficulties is to select a representative group of a very heterogeneous population. In some cases, the choice is intentionally biased towards the healthiest subjects, those who are able and willing to relate details of their past life, complete thorough questionnaires, and undergo special physical and psychological tests. In other studies, the method of recruitment and sometimes the low rate of participation may cast doubt on some kind of unintentional selection. Louhija (1994) had the unique

---

* Centre Claude Bernard de Gérontologie, Hôpital Sainte Périne, 11 rue Chardon-Lagache, 75016 Paris, France

J.-M. Robine et al. (Eds.)
Longevity: To the Limits and Beyond
© Springer-Verlag Berlin Heidelberg New York 1997

opportunity to examine almost all (98 %) of the centenarians living in Finland at the time of his study: only four of the 185 eligible subjects refused to participate. This confers a high degree of reliability on this work; however, it would be unwise to extend its findings without caution to other countries where many important factors (genetic and ethnic characteristics, history, customs, social conditions, health system) are not comparable. In the largest survey of centenarians, which was carried out in 1989 by the French IPSEN Foundation (Allard 1991), medical examinations were done on 756 persons (663 women and 93 men), about a quarter of the estimated total number of centenarians living in France at that time.

A Wide Range of Variations

The differences between studies in selection, methods of investigation and objectives may explain the large differences between groups observed in the literature. The health of institutionalized subjects is usually poorer than that of those living at home (Eeckhoudt 1991; Louhija 1994). However, a very wide range of variation appears between individuals from the same study, provided that the examined sample is representative enough of the population of centenarians. It is well known that inter-individual differences tend to increase with age. This trend does not seem to slow down in late life, and the centenarians probably constitute the most heterogeneous of all the age groups. Past and present way of life, medical history, social status and education vary greatly. The diversity in present physical abilities and cognitive functions is still more striking.

Health of Centenarians: The Pessimistic View

Although there is no "average" centenarian and some of them retain remarkable faculties, the effects of time on general condition are usually serious. Studies that are not intentionally focused on outstanding subjects show a majority of rather severely disabled people.

Cognitive Functions

In the Finnish study, which was nearly exhaustive (Louhija 1994), only one of four centenarians was fully normal according to the Mini D test, and two of five were demented. In a group of 40 centenarians (age range: 100–107) Powell (1994) found common bradyphrenia and bradykinesia and impaired awareness and concern. A common pattern of dementia consisted of preserved awareness of the environment, normal participation in conversations, mild bradyphrenia and bradykinesia with normal latency to respond to questions and memory impairment with diminished ability to learn new information. These patients had a constricted universe with limited awareness of events outside their personal sphere;

they repeated themes and topics endlessly (Powell 1994). The Folstein Mini-Mental State Exam and Washington University's Clinical Dementia Rating Scale indicated moderately advanced dementia in more than half of the subjects; four had a clinical pattern that suggested senile dementia of the Alzheimer's type. In the IPSEN study (Allard 1991), two centenarians out of three had memory impairment, and only one of five passed the Pfeiffer's test with no error. In a group of 121 Hungarian centenarians, the prevalence of dementia was about 50 % (Ivan 1990). In the Georgian study, the performances of psychometric intelligence and memory functions were lower in the centenarians than in younger comparative cohorts (80 and 60 years). Lower performances were also observed in crystallized intelligence and tertiary memory: "the deterioration of performance for the oldest-old seems to be a robust phenomenon in the intelligence domain" (Poon 1992).

Physical Abilities

In the French survey, help was needed by one third of centenarians to eat, by one half to go to the toilet, and by two thirds to get dressed and to get washed. Two thirds were unable to walk 500 meters or to bear a weight of 5 kg; four of five could not climb 10 stairs (Allard 1991). Two thirds had urinary incontinence and one half had fecal incontinence.

One half were confined to bed or armchair; two thirds never left their room. Nearly one half had bad or very bad sight and more than a third had bad or very bad hearing. Only one third could easily watch the TV, and only two of five could listen to the radio. Similar figures appear in the Hungarian study, where over one third of subjects experienced serious loss of vision, and one of six was totally blind from severe degeneration of the macula (Fürjes 1990).

Diseases of Centenarians

Cardiovascular diseases were the most frequently observed in Hungarian centenarians (Gergely 1990), with heart failure in one of four. In the Finnish study, which was nearly exhaustive (Louhija 1994), more than two of three centenarians had a cardiovascular disease, most often congestive heart failure. One half had abnormal ventricular conduction, one of six had atrial fibrillation, and one of five had a defect in atrio-ventricular conduction. ECG findings in Hungarian centenarians were roughly comparable, with only about 10 % of normal records (Rohla and Lengyel 1990). Hypertension was observed only in one of 10 Finnish centenarians, and the mean blood pressure was 142/73 mmHg (versus 137/77 mmHg in the French study). Breast cancer was diagnosed in 4 % of women and prostate cancer in 7 % percent of men. One of five centenarians exhibited a skin cancer, mostly of the baso-cellular type (Louhija 1994; Biro and Regius 1990). Non-insulin-dependent diabetes mellitus was diagnosed in 10 %. Parkinson's disease was surprisingly rare (2 %).

Sex Disparities

Centenarians are Mostly Women

In all developed regions, women outlive men. The difference in life expectancy varies from country to country and is higher in France (around 8 years) than in the majority of other European nations. Therefore it is not surprising to observe a high female to male ratio in the centenarian population. The ratio is 7/1 in France, 5/1 in Finland, and 3.5/1 in Hungary. In these three countries, World Wars I and II killed many young men in the generation of today's centenarians; however, in Sweden, which was a neutral country, the ratio developed in a small survey (52 subjects) is 5.7 (Nordbeck et al. 1991). The predominance of men (323 of 555 cases) in Belle Boone Beard's study is a very unusual sex ratio in a centenarian population. Dr. Beard's editors acknowledge that, as centenarians were identified through public sources and males are more visible in public life, she may have found male case studies more interesting (Beard 1991). One can also think that male centenarians, because their general condition was better, were most likely to be selected in a book that was intended to be optimistic.

Healthy Centenarians are Mostly Men

In all studies, men centenarians are in better condition than women. Men have better abilities, they need less assistance in eating, getting dressed and standing up, they go outdoors more often and take care of the household more often, which is unusual at younger ages. Men have more teeth of their own; their sight and hearing are better. Women tend to use more drugs of any kind, and they have significantly more cardiovascular diseases than men (Beregi et al. 1995).

Impaired memory is more often reported by the family of centenarian women. In the IPSEN study, more men than women were able to be tested with the Pfeiffer's Short Portable Questionnaire, and the mean score of men was significantly better. Dementia is significantly more frequent in women than in men: about one of two vs. one of five. The figures are approximately the same in the French, Hungarian and Finnish studies (Allard 1991; Ivan 1990; Louhija 1994). Men exhibit a better mood and are less often self-centered, they have significantly better language ability, social integration, orientation, sphincter continence, physical independence and mobility than women.

There is no satisfactory explanation for the predominance of women in the centenarian population, nor for the better health state of men. Men seem to be more susceptible to a large variety of pathogens that exert a severe selection, leaving only a small "biological elite" in late life.

Health of Centenarians: The Optimistic View

Over two thirds of French centenarians rated their present quality of life as rather good or very good, although this statement was shared by their relatives in only 55 % of cases. Similarly, four of five considered themselves healthy or fairly healthy, even if only 57 % of their reports were corroborated by relatives.

Although cognitive decline of various degrees can be demonstrated in most centenarians, serial autopsies have revealed that, even in that population, neuropathologists could make a clear distinction between normal aging and dementia. This finding strongly suggests that dementia is not inescapable with very advanced age.

It is certainly fair to admit that many centenarians are more or less severely disabled, but a not insignificant proportion of them can enjoy a life style that could be regarded as normal for much younger persons. For instance, some enjoy tourism. During the year preceding the IPSEN survey, 12 of the centenarians had visited a foreign country and two did so without help. In the same year, 56 had been to another region, three of them without help. During the last three months of the study, 104 had been to another city, 11 without help. During the last 30 days, 29 had used public transportation, 21 of them without help.

Mrs. Jeanne Calment, who celebrated her 121st birthday in February 1996, still rode a bicycle at the end of her first century and she waited until the age of 110 to enter a retirement home. When she was 118, her performance on tests of verbal memory and language fluency was comparable to that of persons with the same level of education in their eighties and nineties (Ritchie 1995). Several French political leaders were still seeking advice from their former colleague, Mr. Antoine Pinay, when he was entering his 101st year. Many other examples of outstandingly well-performing centenarians are reported in the literature (Zhengcai 1990; Allard 1991; Beard 1991). Such cases are obviously quoted because they seem quite exceptional; however, besides everyday geriatric experience, several arguments militate in favor of a progressive improvement in newer centenarians' health. While the number of centenarians increases rapidly and that increase in likely to continue (Vaupel and Gowan 1986), their mortality is declining (Kannisto 1993). Moreover, several functions exhibit improvement at the same age in successive cohorts, and the progression of some expressions of the aging process seems to slow down in younger cohorts compared to their elders (Svanborg et al. 1984).

Centenarians are more and more eligible to take advantage of medical and surgical progress. From a geriatric point of view, two major breakthroughs have been achieved in surgery during the last decades: joint and lens replacement. Joint prosthesis has radically transformed the prognosis of hip fractures, allowing very old patients to resume walking within a few days instead of remaining bedridden forever when they did not die rapidly from decubitus infections or ulcers. Lens replacement does not represent such a vital necessity, but the improvement in quality of life it gives to patients after cataract surgery is incomparable. Centenarians are no longer excluded on the sole basis of age, provided their general condition is adequate. The same accessibility applies to artificial cardiac pacemaker implantation and other techniques such as new, non-invasive methods of investigation or surgery that were reserved for younger patients some years ago. Ethical limits are rather easy to draw with a minimum of common sense, but the major obstacles will certainly remain economic considerations and prejudice against old people ("agism").

Better living conditions and medical progress will probably increase the number of centenarians in the next future. How will their average state of health be if these changes allow persons who do not have the strong genetic background of today's centenarians to live 100 years and more? Are we moving towards a "compression of morbidity" (Fries 1980) or will the number of disabled and dependent elderly people become increasingly overwhelming (Schneider and Brody 1983)? We have seen several reasons to believe that centenarians will be better in future generations if the present trend merely persists. But the aging process itself does not seem to be out of reach of gerontological research (Christen 1991). The day we can slow down the biological clock, centenarians will become very ordinary persons.

Summary

Studies on centenarians are still relatively scarce, few studies include more than 100 persons, and in some cases the choice is intentionally biased towards the healthiest subjects. The differences in selection, methods of investigation and objectives may explain the large variations between groups observed in the literature. Although there is no "average" centenarian and some of them retain remarkable faculties, the effects of time on their general condition are usually serious. Studies that are not intentionally focused on outstanding subjects show a majority of rather severely disabled people, with alterations in cognitive functions and restricted physical ability. Cardiovascular diseases are also frequent. As women outlive men, it is not surprising to observe a high female to male ratio in the centenarian population (7/1 in France). In all studies, male centenarians are in better condition than female centenarians, they retain better capacities and they need less assistance for the activities of daily living. Over two thirds of the French centenarians rated their present quality of life as rather good or very good, although this statement was not always shared by their relatives. Several arguments militate in favor of a progressive improvement in the health of future centenarians. Some functions exhibit improvement at the same age in successive cohorts, and the progression of some expressions of the aging process seems to slow down in younger cohorts compared to their elders. Centenarians are more and more eligible to take advantage of medical and surgical progress. Ethical limits are rather easy to draw with a minimum of common sense, but the major obstacles will certainly remain economic considerations and prejudice against old people.

References

Allard M (1991) A la recherche du secret des centenaires. Le Cherche-Midi, Paris

Bancher C, Jellinger KA (1994) Neurofibrillary tangle predominant form of senile dementia of Alzheimer type: A rare subtype in very old subjects. Acta Neuropathol 88:565–570

Barbagallo CM, Averna MR, Frada G, Chessari S, Mangiacavallo G, Notarbartolo A (1995) Plasma lipid, apolipoprotein and Lp(a) levels in elderly normolipidemic women: relationships with coronary heart disease and longevity. Gerontology 41:260–266

Beard BB (1991) Centenarians, the new generation. Wilson NK, Wilson AJE III (eds) Greenwood Press, New York

Beregi E (1990) Centenarians in Hungary. A social and demographic study. Interdisciplinary topics in gerontology. Vol 27. Karger, Basel

Beregi E, Regius O, Nemeth J, Rajczy K, Gergely I, Lengyel E (1995) Gender differences in age-related physiological changes and some diseases. Z Gerontol Geriatr 28:62–66

Biro J, Regius O (1990) Dermatological studies in centenarians. In: Beregi E (ed) Centenarians in Hungary. A social and demographic study. Interdisciplinary topics in gerontology. Vol 27. Karger, Basel, pp 83–85

Christen Y (1991) Les années Faust ou la science face au vieillissement. Sand, Paris

Clarke CA, Mittwoch U (1994) Puzzles in longevity. Perspectives Biol Med 37:327–336

Cobler JL, Akiyama T, Murphy GW (1989) Permanent pacemakers in centenarians. J Am Geriatr Soc 37:753–756

Delaere P, He Y, Fayet G, Duykaerts C, Hauw JJ (1993) A-4 deposits are constant in the brain of the oldest old: an immunocytochemical study of 20 French centenarians. Neurobiol Aging 14:191–194

Eeckhoudt A (1991) Caractéristiques de treize centenaires parisiennes. Université Paris VI Faculté Broussais-Hotel-Dieu

Effros RB, Boucher N, Porter V, Zhu XM, Spaulding C, Walford RL, Kronenberg M, Cohen D, Schachter F (1994) Decline in CD28(+) T cells in centenarians and in long-term T cell cultures: A possible cause for both in vivo and in vitro immunosenescence. Exp Gerontol 29:601–609

Fayet G, Hauw JJ, Delaere P, He Y, Duykaerts C, Beck H, Forette F, Gallinari C, Laurent M, Moulias R, Piette F, Sachet A (1994) Neuropathologie de 20 centenaires. I Données cliniques. Rev Neurol 150:16–21

Franceschi C, Monti D, Sansoni P, Cossarizza A (1995) The immunology of exceptional individuals: the lesson of centenarians. Immunol Today 16:12–16

Fries JF (1980) Aging, natural death, and the compression of morbidity. N Engl J Med 303:130–135

Fürjes E (1990) Ophthalmological examinations of centenarians. In: Beregi E (ed) Centenarians in Hungary. A social and demographic study. Interdisciplinary topics in gerontology. Vol 27. Karger, Basel, pp 65–82

Gergely I (1990) The state of health of centenarians from the point of view of internal medicine. In: Beregi E (ed) Centenarians in Hungary. A social and demographic study. Interdisciplinary topics in gerontology. Vol 27. Karger, Basel, pp 40–46

Giannakopoulos P, Hof PR, Vallet PG, Giannakopoulos AS, Charnay Y, Bouras C (1995) Quantitative analysis of neuropathologic changes in the cerebral cortex of centenarians. Progr Neuro-Psychopharmacol Biol Psychiat 19:577–592

Hauw JJ, Vignolo P, Duykaerts C, Beck M, Forette F, Henry JF, Laurent M, Piette F, Sachet A, Berthaux P (1986) Etude neuropathologique de 12 centenaires: La fréquence de la démence sénile de type Alzheimer n'est pas particulièrement élevée dans ce groupe de personnes très âgées. Rev Neurol 142:107–115

Hauw JJ, Delaere P, Fayet G, He Y, Costa C, Seilhean D, Duykaerts C, Beck H, Forette F, Gallinari C, Laurent M, Moulias R, Piette F, Sachet A (1993) The centenarian's brain. Sandoz, pp 17–25

Ivan L (1990) Neuropsychiatric examination of centenarians. In: Beregi E (ed) Centenarians in Hungary. A social and demographic study. Interdisciplinary topics in gerontology. Vol 27. Karger, Basel, pp 53–64

Karasawa A (1979) Mental aging and its medico-social background in the very old Japanese. J Gerontol 34:680–686

Klatt EC, Meyer PR (1987) Geriatric autopsy pathology in centenarians. Arch Pathol Lab Med 111:367–369

Kannisto V (1993) La mortalité des centenaires en baisse. Population 4:1070–1072

Kinzel T, Wekstein D, Kirkpatrick C (1986) A social and clinical evaluation of centenarians. Exp Aging Res 12:173–176

Louhija (1994) Finnish centenarians. Academic dissertation, University of Helsinki

Mari D, Mannucci PM, Coppola R, Bottasso B, Bauer KA, Rosenberg RD (1995) Hypercoagulability in centenarians: the paradox of successful aging. Blood 85:3144–3149

Mariotti S, Barbesino G, Caturegli P, Bartalena L, Sansoni P, Fagnoni F, Monti D, Fagiolo U, Franceschi C, Pinchera A (1993) Complex alteration of thyroid function in healthy centenarians. J Clin Endocrinol Metab 77:1130–1134

McCann WJ, Smith JW (1990) The surgical care of centenarians. Curr Surg 47:2–3

Nordbeck B, Alfredson B, Hagberg B, Samuelsson G, Samuelsson SM (1991) The Swedish centenarian study: quality of life. II European Congress of Gerontology, Madrid (poster)

Paolisso G, Gambardella A, Balbi V, Ammendola S, Damore A, Varricchio M (1995) Body composition, body fat distribution and resting metabolic rate in healthy centenarians. Am J Clin Nutrit 62:746–750

Poon LW (1992) The Georgian Centenarian Study. Baywood, Amytyville, New York

Powell AL (1994) Senile dementia of extreme aging – a common disorder of centenarians. Dementia 5:106–109

Ritchie K (1995) Mental status examination of an exceptional case of longevity. J C aged 118 years. Br J Psychiat 166:229–235

Rohla M, Lengyel E (1990) Studies of centenarians in Hungary: results of twelve-lead electrocardiographic analysis. In: Beregi E (ed) Centenarians in Hungary. A social and demographic study. Interdisciplinary topics in gerontology. Vol 27. Karger, Basel, pp 47–52

Schneider EL, Brody J (1983) Aging, natural death, and the compression of morbidity: another view. N Engl J Med 309:854–856

Sobel E, Louhija J, Sulkava R, Davanipour Z, Kontula K, Miettinen H, Tikkanen M, Kainulenen K, Tilvis R (1995) Lack of association of apolipoprotein E allele epsilon 4 with late-onset Alzheimer's disease among Finnish centenarians. Neurology 45:903–907

Svanborg A, Berg S, Nilsson L, Persson G (1984) A cohort comparison of functional ability and mental disorders in two representative samples of 70-year-olds. In: Wertheimer J, Marois M (eds) Senile dementia: New York, Alan R. Liss, pp. 405–409

Takata H, Suzuki M, Ishii T, Sekiguchi S, Iri H (1987) Influence of major histocompatibility complex region genes on human longevity among Okinawan-Japanese centenarians and nonagenarians. Lancet ii:824–826

Vaupel JW, Gowan AE (1986) Passage to Methuselah: some demographic consequences of continued progress against mortality. Am J Public Health 76:430–433

Zhengcai L (1990) Les secrets de la longévité. Editions en langues étrangères. Beijing, China

Looking Into the Crystal Ball: Will We Ever be Able to Accurately Predict Individual Differences in Longevity?

L. W. Poon, M. A. Johnson[*] and P. Martin[**]

The purpose of this chapter is to highlight issues in the study of successful aging on the one hand and predicting longevity on the other. Intuitively, one would ask: if a person ages successfully doesn't that person have a higher probability of living longer? The answer to this question is both YES and NO. The key question that needs to be answered is how different combinations of factors involving nature and nurture combine to contribute to individual differences in longevity. The implication of this question is related to predetermination of one's life span on the one hand and self and other determinations on the other.

Our current state of knowledge on aging precludes predicting longevity at an individual level. One of the lessons learned from longitudinal studies of aging is that few individuals follow the pattern of age changes described in average longitudinal functions (e.g., Dannefer and Sell 1988; Pedersen and Harris 1990). A similar phenomenon exists in cross-sectional studies of aging (e.g., Siegler et al. 1995). Averages that are supposed to describe composite characteristics of sample groups often are not representative of any one individual. This is a difficult state of affairs in predicting longevity on an individual basis. It is particularly a problem in predicting longevity of the very old, as both intra- and inter-individual variability of some functions increase in advanced age.

Successful Aging

Rowe and Kahn (1987) wrote an influential paper that was published in Science titled, "Human Aging: Usual and Successful." This is an often cited paper that is important in our understanding of successful aging and predicting longevity.

This paper made four important points:
1) Research in aging has emphasized age-related loss.
2) The effects of the aging processes have been exaggerated.
3) The modifying effects of extrinsic factors such as diet, exercise, personal habits, and psycho-social factors on aging have been underestimated.

---
[*] University of Georgia, Athens, Georgia
[**] Iowa State University, Ames, Iowa, USA

J.-M. Robine et al. (Eds.)
Longevity: To the Limits and Beyond
© Springer-Verlag Berlin Heidelberg New York 1997

4) The substantial individual differences within groups has either been ignored or attributed to differences in genetic makeup.

Rowe and Kahn's main points addressed the limitation of our current understanding of the aging processes. Specifically, they addressed our current lack of understanding of the interplay within and across bio-psycho-social factors as they relate to aging. They discussed the possibility that some of these bio-psycho-social factors may have facilitative as well as inhibitory effects on the quantity and quality of life. They also admonished us for believing too much in genetic determinism passed on by our ancestors.

Rowe and Kahn used these postulations to propose and differentiate two concepts. They first identify a pattern of aging called usual aging, in which extrinsic factors such as life style, diet, support system, and psycho-social factors tend to exacerbate the effects of aging. This was differentiated from successful aging, in which extrinsic factors either have minimal or positive effects.

The research described here is a study of age-related differences in "naming latency" (Poon and Fozard 1978) to illustrate Rowe and Kahn's assertion of the effects of extrinsic factors on the understanding of aging processes. The speed of naming objects, or naming latency, has implications about the efficiency of the neuropathways of an individual. The question is whether older people have demonstrable losses of this ability. Young, middle aged, and older subjects were presented with four types of pictures through a computer-controlled tachistoscope. A voice-activated relay was used to time the naming latency between the onset of the picture and the onset of the voice response. The four types of pictures were exemplars of objects that were used in the 1980s or in the 1920s. Half of the objects were commonly used objects such as shoes or combs. The other half of the objects were equally divided between those that were used primarily in the 1920s (e.g., churn) and in the 1980's (e.g., digital clock). Only the naming latency of correctly identified objects was used in the analyses.

Figure 1 shows the naming latency of the young, middle aged, and older subjects for the four types of objects. There were hardly any differences in the naming latency of frequently used objects in the three age groups. The constant difference among the three age groups does reflect the slowing of psychomotor speech due to aging. The correct naming latency of those objects used in the 1920s and 1980s presented interesting patterns. First look at the naming latency of objects used in the 1980s. The latency was fastest for the young, followed by the middle aged and the older subjects. One would recognize this as a classic aging function. Examining this function alone, one could come to a conclusion that there is systematic age-related loss in the speed of correctly naming objects. However, the story does not end here. Examining the naming latency of objects used in the 1920s, we have an almost reversed aging function. Younger subjects performed more poorly than the older adults.

There could be a number of different explanations for the results. In the examination of the naming latency function of objects used in the 1920s, in which younger subjects were slower than older adults, most researchers would

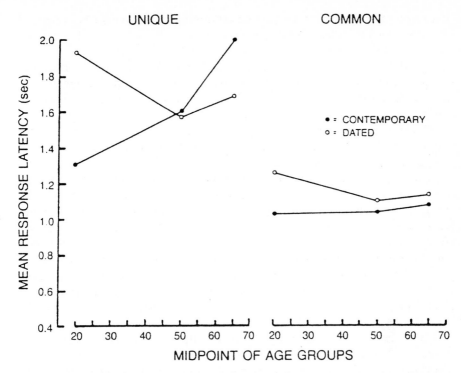

Fig. 1. Mean naming latencies for the unique and common versions of contemporary and dated exemplars for the 20-, 50-, and 65-year-old groups. (Reprinted with permission from Poon and Fozard 1978).

not say that the younger subjects were aging faster than the older individuals. However, when presented with the latency function of objects used in the 1980s, in which older adults were slower, most would interpret it as support of differential loss due to age. Some may even offer hypotheses of differential atrophy of the central neurologic or peripheral visual or speech pathways due to aging. The explanation that was most cogent is that age-related familiarity with the stimuli could exert a profound effect on the naming latency of the subject regardless of chronological age.

This was a clear demonstration that a psychological extrinsic factor of age-related familiarity with the object to be named could either increase or decrease the efficiency of performance for individuals of different ages. More importantly, this was a demonstration that a better understanding of an extrinsic effect could keep us from drawing a wrong conclusion on an aging process. Perhaps we need to be reminded that we do not yet know enough about the contributions of individual bio-psycho-social factors that may directly, indirectly, or in combinations contribute to the aging processes.

According to the explications of Rowe and Kahn, successful aging is found where extrinsic factors either do not contribute to the negative effects of aging or slow down the effects of aging. Two logical questions are whether and how successful aging contributes to longevity in general on the one hand, and to individual differences in longevity on the other hand.

Given this conceptual framework, the search for sensitive predictors of successful aging is a complex task. We have only begun to untangle the web of complexity. At least six levels of complexity need to be addressed in the study of successful aging.

One, the outcome criteria of successful aging need to be defined, and they may be different for different purposes. One may define success by its quality or quantity or a combination thereof.

Two, the selection of appropriate constructs that may best predict the outcome is a challenge. These predictor constructs would need to include the extrinsic factors posed by Rowe and Kahn as well as other combinations of bio-psycho-social factors. It is interesting to note that there is general consensus about predictors in the literature. The following are some generally agreed upon domains necessary for predicting successful aging: genetic contribution, length of life, biological health, mental health, cognitive efficacy, personality, social competence and productivity, personal control, and life satisfaction (Baltes and Baltes 1990). It is interesting to note that these constructs could be employed both as predictor and criterion variables.

Three, the selection of appropriate indicators that could best measure a construct is controversial at best. For example, if cognitive competency is an important predictor, which cognitive measures should be employed to estimate competence? Should there be general or specific estimates of cognitive processes? If genetic contribution needs to be included, which genetic markers are most sensitive? If biological health needs to be included, which biological functions should be measured?

Four, given that one employs a multivariate approach in predicting successful aging, how should one conceptualize the direct and indirect influences of these predictors in order to realize the maximum predictive power? The theoretical model one would select to predict successful aging would ultimately determine the efficacy of the effort.

Five, depending on the definition of successful aging, one must be prepared for the possibility that different predictors or combinations of predictors may be salient for different stages of the life span.

Finally, if sufficient samples of individuals are obtained within a specific cohort, then it would be important to identify subtypes of the sample where specific predictors and combinations of predictors may be more salient in predicting successful aging. This may be where sensitivity to individual differences could pay off in increasing the accuracy of prediction.

Successful Aging and Longevity

Longevity is a univariate number describing the length of life. Successful aging or successful adaptation, on the other hand, defines a set of criteria that allow a person to attain a high quality of life. Measures of successful aging need to capture the multitude of ingredients that would meet the predefined criteria. Figure 2 is an example of a multivariate model measuring successful adaptation by the Georgia Centenarian Study (Poon et al. 1992). The study examines the contributions of family longevity, support systems, individual adaptation skills, nutrition, and physical and mental health to life satisfaction and morale in community-dwelling and cognitively intact individuals aged in their 60s, 80s, and 100-plus years.

An important question in predicting successful aging and longevity at the individual or group level is whether similar predictors are pertinent for both. Although there is no answer to that question at present, one might speculate that many similar predictors would be pertinent. Figure 3 shows an example of the Bonn Longitudinal Study (Thomae 1976) in which similar predictors were employed for longevity.

Extension of the designs and findings from these two types of studies has the potential of providing us with an understanding of successful aging and longevity. Extant findings seem to suggest that different predictors and different

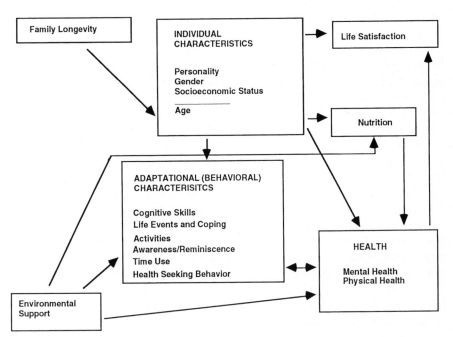

Fig. 2. Georgia Centenarian Study general conceptual model. (Adapted with permission from Poon et al. 1992).

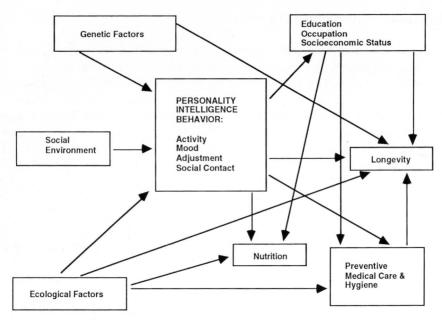

Fig. 3. Bonn theoretical model. (Adapted with permission from Poon et al. 1992).

weightings of the same predictors are pertinent for individuals at different ages for life satisfaction. Similarly, different weighting of similar predictors may be found for the prediction of life satisfaction and longevity. Further, it may be possible to identify specific predictors that seem to particularly influence subtypes of individuals in both life satisfaction and longevity. For example, some centenarians tend to be the heirs of generations of long-lived individuals, including a number of centenarians in the family (Woodruff-Pak 1988). This pattern was found in only a minority of centenarians in the Georgia Centenarian Study (Poon et al. 1992).

We found that community-dwelling and cognitively intact centenarians consume 21% to 30% more vitamin A and carotene than those in their 60s or 80s (Fischer et al. 1995). We also found that these centenarians tend to be more dominant and suspicious but less tense and at the same time centenarians tend to be able to draw on cognitive coping resources that would allow them to reflect and think through what they find troublesome in their lives (Martin et al. 1992). Are these examples of special characteristics typical for all long-lived individuals? Were these life-long characteristics or were they acquired over time? How well could those extrinsic factors pertinent to successful aging (Rowe and Kahn 1987) predict longevity at the group and individual levels? The search for these and other answers promises to be exciting, particularly with an increased emphasis on bio-psycho-social collaboration.

References

Baltes PB, Baltes MM (1990) Psychological perspectives on successful aging: The model of selective optimization with compensation. In: Baltes PB, Baltes MM (eds) Successful aging: Perspectives from the behavioral sciences. New York, Cambridge University Press

Dannefer D, Sell RR (1988) Age structure, the life course and "aged heterogeneity": Prospects for research and theory. Comp Gerontol B 2:1–10

Fischer JG, Johnson MA, Poon LW, Martin P (1995) Dairy product intake of the oldest old. J. Am Diet Assoc 95:18–21

Martin P, Poon LW, Clayton GM, Lee HS, Fulks JS, Johnson MA (1992). Personality, life events, and coping in the oldest-old. Int J Aging Human Devel 34(1):19–30

Pedersen NL, Harris JR (1990) Developmental behavioral genetics and successful aging. In: Baltes PB, Baltes MM (eds) Successful aging: Perspectives from the behavioral sciences. New York, Cambridge University Press

Poon LW, Fozard JL (1978) Speed of retrieval from long-term memory in relation to age, familiarity, and datedness of information. J Gerontol 33:711–717

Poon LW, Clayton GM, Martin P, Johnson MA, Courtenay BC, Sweaney AL, Merriam SB, Pless BS, Thielman SB (1992) The Georgia Centenarian Study. Int J Aging Human Devel 34(1):1–18

Rowe JW, Kahn RL (1987) Human aging: usual and successful. Science 237:143–149

Siegler IC, Poon LW, Madden DJ, Welsh KA (1995) Psychological aspects of normal aging. In: Busse EW, Blazer DA (eds) Geriatric psychiatry Washington, DC, American Psychiatric Press, pp. 105–128

Thomae H (1976) Patterns of aging: Findings from the Bonn longitudinal study of aging. New York, Karger

Woodruff-Pak DS (1988) Psychology and aging. Englewood Cliffs, NJ, Prentice-Hall, Inc

Towards a Genealogical Epidemiology of Longevity

J.M. Robine [*] *and M. Allard* [**]

Like father, like son? What is the relation between the parents' or grandparents' longevity and that of their children? To what extent does genetic endowment influence the differences in longevity observed between individuals?

What is the effect of the selection brought about by infant mortality or early mortality on the survivors' robustness and longevity at different ages of life? What effect do paternal or maternal ages at conception have on their children's robustness and longevity? What effect does late procreation have on both the parents' and the children's longevity?

All of these basic questions about human longevity have been asked for a long time but answers have not yet been found. Thus Gavrilov and his colleagues in 1996 are asking almost the same questions regarding the possible role of paternal age at the child's conception as Buffon did in 1749. By the end of 1995, when they presented their work at the IPSEN Foundation, Gavrilov et al. indicated that some Russian genealogical data suggested that paternal age at a child's conception had a pejorative effect on the child's life duration. They added that a thorough review of the literature allowed them to assert that this question had never been really studied.

In fact, the first studies that were carried out regarding these questions in the eighteenth and nineteenth centuries dealt with the effect of paternal age.

Thus, according to Buffon (1749), for the seminal fluid to be prolific, it must contain organic molecules from all the parts of the body, which is not the case in old men whose bones and other body parts have become too solid. According to Buffon, "we should think that these missing molecules may sometimes be replaced by those of the female, if she is young." However that may be, senile old men seldom procreate and, when they do, "they often engender monsters, deformed children, still more defective than their father."

Later, at the beginning of the twentieth century, it was the heritability of longevity that was investigated. The most famous works are those of Beeton and Pearson (1901), based on some English genealogies, which were published at the turn of the century, and especially those of Pearl, based on an important sample of nonagenarians, which were published in the 1930s.

[*] INSERM
[**] IPSEN Foundation

J.-M. Robine et al. (Eds.)
Longevity: To the Limits and Beyond
© Springer-Verlag Berlin Heidelberg New York 1997

Raymond Pearl proposed to calculate the "Total Immediate Ancestral Longevity" (abbreviated TIAL, pronounced tee-aal), that is the sum of the ages at death of the six immediate ancestors of an individual (Pearl and Pearl Dewitt 1934).

These studies and those carried out later concluded that a strong heritability of longevity existed. The study designs used to examine these questions were contestable – such as, for instance, the absence of an adequate validation of life durations. They were severely criticized by Murphy in 1978 and by Jacquard in 1982. The latter argued that individual differences in life durations are not genetic in origin.

Finally, more recently, research has been carried out on the effect of a decrease in selection forces.

Compared with the works carried out on animals, *Drosophila Melanogaster* or rodents, only a very few studies have been devoted recently to the longevity of the human species.

For example, Pierre Philippe (1980) has studies the relations between maternal age at birth and children's life durations in the Isle-aux-Coudres in Quebec. He observed a link between maternal age and life durations that were less than five years or greater than 70 years for the cohort that was constituted between 1880 and 1899. He noted that there was no link with paternal age when the maternal age was known. Even if a high maternal age was associated with a life duration superior to 70 in most cohorts, he did not find in the 1800–1879 cohorts, when the under-declaration of infant mortality was more important, the results that had been observed in the 1880–1899 cohort.

In 1990, Desjardins and Charbonneau confirmed from the population register of old Quebec that, in the seventeenth and eighteenth centuries a family component undeniably existed in longevity, although it was rather weak. When the children died at higher ages, the parents' average ages at death were higher. There was some homegeneity in brothers' and sisters' life durations, whereas there was no correlation between spouses' life durations. They remarked that "the absence of any link does not mean, for all that, that the milieu has no significant effect in this case, but this absence throws at least some light on the intergenerational correlations observed."

In 1991, Bocquet and his collaborators concluded from a study of the population of Arhez d'Asson focused on the cohorts from 1744 to 1850 that the genetic influence on a individual's potential longevity was very weak. However, they noted that this influence increased in the course of life. It could reflect a more important transmissibility of degenerative diseases beyond age 60. They emphasized that maternal mortality could account for the obscure paternal influences on longevity that some researches thought they had detected from the established statement that the father-children correlations were more important. They noted that there was no effect of social homogamy between spouses, from which they concluded that "professional categories" were not synonymous with "social stratifications" at Arthez d'Asson (!). Finally, they pointed out the difficulties in this type of studies which are raised by the secular increase of age at death.

Also in 1991 Kannisto studied the effect of external factors (like epidemics) on the survivors' frailty or vigor, from the Finnish historical series. He pointed out that only proximate cohorts can be compared with one another, as they have been subject to approximately the same period effects. And "since the genetic heritage of population changes only very slowly, contemporary cohorts can be considered to have virtually equal degree of frailty at conception. They are affected only by individual differences in parent generation and the probability of heredity." He noted that cohorts remained very vulnerable to external influences at birth and during childhood. On the one hand, epidemics may leave the more robust alive and, on the other hand, they may leave the survivors more frail. For Kannisto, indeed, "it would also be logical to expect the same external factors to impair the health of survivors." He concluded that genetic frailty appears to be very weak and that, if an external event may significantly increase the period mortality, survivors are neither more robust, as the natural selection theory pretends, nor are they made durably more frail.

In 1993, Le Bourg and his colleagues studied the possible existence of a trade-off between early fecundity and longevity – also using the population registers of old Quebec – in a noncontraceptive human population living at a time when longevity had not been prolonged by medical care and was not artificially shortened by wars, epidemics, or other external causes. The aim of the study was to test if a trade-off between early fecundity and longevity does exist in women living under a natural fertility regimen. According to the Williams theory, "It could be supposed that women who give birth to high numbers of children in early life would have a short life-span." Results: on the whole, the most long-lived Canadians gave birth to a higher number of children and had a higher early fecundity.

The authors concluded: "Yet, if it was impossible to show (by computing phenotypic correlation) that in a heterogeneous population, individuals with high early fecundity and low longevity can be opposed to individuals with low early fecundity and high longevity, it could be concluded that genes coding for that have no actual importance in the variability of longevity in these populations."

To go further in the study of the heritability of longevity, in 1993, Matt Mc Gue, James Vaupel and their colleagues studied Danish twins born between 1870 and 1888. They noted that, although a family component in longevity existed – however modest – parental age at death appeared to have minimal prognostic significance for offspring longevity. Familial resemblance may be the result of shared environment as well as shared genes. They clearly distinguished between genetic factors, shared environment and non-shared environment, i.e., those that contribute to dissimilarity. They pointed out that the difference in age at death for monozygotic, but not like-sex dizygotic, twins was significantly smaller than the difference between two unrelated, like-sex individuals randomly selected. They concluded that there existed a moderate but statistically significant influence of genetic factors on length of life. Environmental factors accounted for a majority of the variance in life duration, but the relevant environmental factors appeared to be the non-shared ones.

On the whole, only a few studies have been carried out on the topic and the available results provide only a few certainties. The only consensus seems to bear on the existence of a family component in longevity, qualified "weak" or "modest". In reality, each study brings up a new element to be considered:

- the variations in the under-declaration of infant mortality,
- the distinction between the shared genes and the shared environment in these family component,
- maternal mortality,
- the secular increase in age at death,
- the proximity of cohorts (so that the cohorts studied have been subject to approximately the same period effects),
- infant and childhood mortality,
- external factors, epidemics,
- the distinction between the shared and the non-shared environment.

Even if the genetic endowment, parental ages at conception, or the selection at the beginning of life have an effect on the variations of individuals' life durations, the influence of those factors is not very strong compared to environmental factors. The links considered here are very tenuous; therefore they are more difficult to point out.

The Archives of Arles

All of the questions raised here could be found in the archives of Arles that we examined thoroughly on the occasion of the validation of the age of Jeanne Calment and the reconstitution of her genealogical tree (Robine and Allard 1995). Although François I ordained that complete patronym and birth hour and day should be written down in registers as early as 1539, it was only in 1922 that mentioning the parents' birth dates and places became compulsory (Act of 1922, October 28th), and only in 1945 that mentioning the death date in the margin of these acts became compulsory (Act of 1945, March 29th). But in Arles, registrars had begun to inscribe the parental ages and reported deaths in marginal notes long before it became compulsory. The only problem left was to put the data on computer. The usefulness of such a project was obvious – considering the rarity of the available empirical data dealing with these questions and the high quality of the data found in Arles – but its estimated cost was considerable. Therefore we decided to start an exploratory study.

There is a paradox between this absence of simple relations between parents' and children's longevity and the recent discoveries concerning the genetic susceptibility to various diseases. For many researchers, the conditions experienced by the cohorts up to now have largely overruled the possible relations between genetic endowment and longevity, making the fraction of the variance of life duration attributable to genetic factors negligible. Along the improvement of living conditions that has resulted in an increase in life expectancy, this attributable fraction could become more important, to the extent that, in ideal environmental conditions and life habits, only genetic differences could still account for the variations in individual life durations.

Thus, we decided to carry out the study cohort after cohort, and to begin with the most recent – that is the cohort born in 1895 – and then to go back to the oldest. In addition to all the life durations per gender up to 1995, the first cohort provided us with the parental age at the children's birth. Through successive additions, the protocol should enable us to 1) compare the proximate cohorts, 2) evaluate the impact of infant mortality on the whole distribution of life durations, 3) estimate the relations between parental age at birth and the children's life durations with respect to the initial level of infant mortality, 4) take into account the secular tendencies and measure the effect of the great mortality crises; then, year after year, 5) reconstitute the siblings and take into account the shared environment, 6) reconstitute the parents' and the grandparents' life durations, 7) and then evaluate the relations between the life durations of children, parents or grandparents, or even 8) evaluate the relations between early fecundity and the life duration of one generation or the other, the other factors being under control.

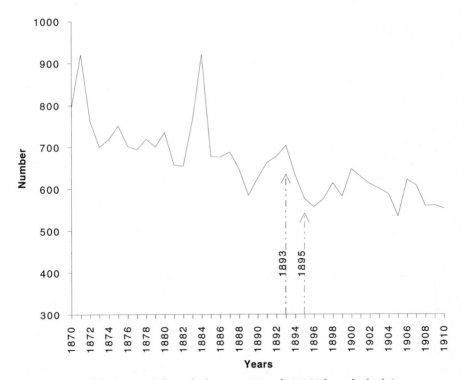

Fig. 1. Number of deaths recorded in Arles between 1870 and 1910 (Arles and suburbs)

Results

The exploratory study has just begun. It involves the first most recent 20 cohorts, ranging from 1895 to 1875. The 1875 cohort comprises Jeanne Calment, who is now 122 years old. In 1884, Arles suffered its last great cholera epidemics (Fig. 1). About 600 children are expected every year. As an example of the problems raised, we are presenting some results from the first three cohorts analysed, successively 1894, 1893 and 1895, because the 1895 registers were not yet available when the analysis started (Table 1).

1894: The analysis of the 1894 cohort (635 children) shows first a strong positive relation between infant mortality and parental age. When the maternal age is known, there is no longer a relation with the paternal age. In women over 30 (around half the mothers), infant mortality is twice as high as in younger woman.

Secondly, in the children surviving beyond 30, there is a clear negative correlation between their own life durations and the paternal age. In the survivors whose fathers were over 35 at their birth (around half the fathers), life expectancy at 30 is reduced by more than 3 years, compared to those whose fathers were younger.

1893: The analysis of the 1893 cohort (616 children) does not confirm any of these results. There is no longer any relation between parental age and infant mortality or between parental age and exact life durations beyond 30.

Infant mortality is very different in the two cohorts: 20.4% in 1893 versus 12.6% in 1894. We can wonder if the relations between parental age and children's life durations are only discernible in the case of low infant mortality.

Moreover, the survivors at 30 in the 1893 cohort – which suffered the highest infant mortality – have a life expectancy at 30 that is five years higher than that of the survivors at the same age in the 1894 cohort.

Table 1. Some characteristics of the 1894, 1893 and 1895 birth cohorts in Arles, France

	1894 cohort	1893 cohort	Comparison 1984:1893	1895 cohort	Comparison 1893:94:95
Number of children	635	616		586	
Complete data (%)	61	62		63	NS
Maternal age (yr)	28.6	28.4		28.1	NS
Paternal age (yr)	34.8	34.4		34.5	NS
Infant Mortality (%)	12.6	20.4	p 0.01	17.9	p 0.01
Mother < 30	8.6	24.4		17.2	
Mother ≥ 30	18.0	17.4		19.0	
	p 0.008	NS		NS	
Life expectancy at 30 (yr)	42.3	47.6	p 0.0001	45.6	p 0.0001
Father < 35	43.8	47.7		44.0	
Father ≥ 35	40.5	47.6		47.6	
	p 0.03	NS		p 0.03	

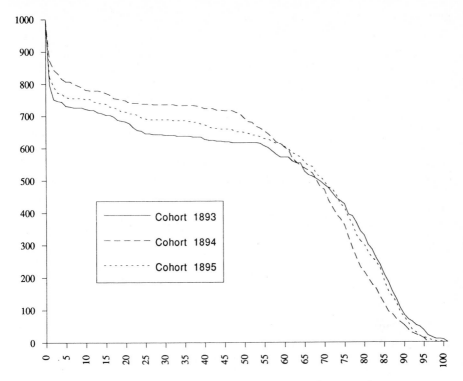

Fig. 2. Survival curves for the three birth cohorts from 1893, 1894 and 1895 in Arles

A comparison of the survival curves (Fig. 2) shows that the number of survivors is 10 % higher in the 1894 cohort between age 25 and age 50, but at age 80, the contrary occurs. The distribution of life durations (Fig. 3) clearly shows that there is an excess mortality between age 45 and age 75 in the 1893 cohort, which suffered the lowest infant mortality, compared to the 1894 cohort.

1895: The analysis of the 1895 cohort (586 children) confirms these results, as this cohort occupies an intermediary position as regards all the criteria: infant mortality, life expectancy at age 30, survivors at age 80 (Table 1, Fig. 1 and 2).

The comparison of the three cohorts in relation with all other criteria – such as the percentage of birth records comprising the father's and mother's ages and the individual's death rate in the margin, the mean paternal and maternal ages at birth – does not show any difference (Table 1), confirming again the high quality of the data.

However, the analysis of the 1895 cohort does not present any relation between infant mortality and parental age, but it shows – regarding the children surviving beyond 30 – a clear, positive correlation between their life duration and the paternal age of their father. In the survivors whose fathers were over 35 at their birth (around half the fathers), life expectancy is increased by 3 years compared to those whose fathers were younger.

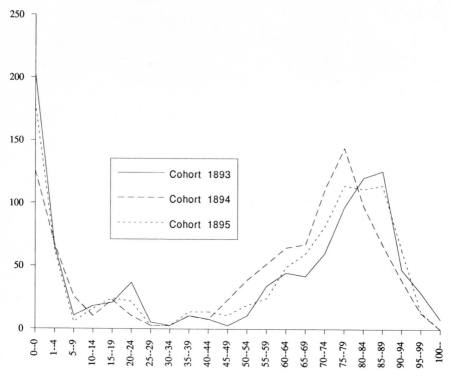

Fig. 3. Distribution of life durations for the three birth cohorts from 1893, 1894 and 1895 in Arles (per 1000)

We cannot currently explain these results, but it is clear that the merging of the three cohorts, to increase the size of the sample (or the power of the tests), could only, in these conditions, result in mitigated, mean results that would not at all correspond to the actual experience of the different cohorts. The results for 1894 confirm the hypothesis put forward by Gavrilov and his colleagues, but those for 1895 contradict it. Kannisto pointed out that only proximate cohorts can be compared with one another. As they have been subject to approximately the same period effects, they are affected only by individual differences in parent generation. This is really intriguing and there are currently too few data of high quality available for us to be capable of drawing a conclusion. Considering the great interest in the topic, we think it is important to gather such data now, carefully avoiding erasing the possible variations between the proximate cohorts or territories.

References

Beeton M, Pearson K (1901) On the inheritance of the duration of life and the intensity of natural selection in man. Biometrika 1:50–89

Bocquet-Appel JP, Jakobi L (1991) La transmission familiale de la longévité à Arthez d'Asson (1685–1975). Population 46:327–347

Buffon (1826) Oeuvres complètes de Buffon. volume 4. Paris, P Duménil (Original edition 1749)

Desjardins B, Charbonneau H (1990) L'héritabilité de la longévité. Population 45(3):603–615

François I (1539) Ordonnance de Villers Cotterêts

Gavrilov LA, Gavrilova NS, Evdokushkina GN, Kushnareva YE, Semyonova VG, Gavrilova AL, Lapshin EV, Evdokushkina NN (1996) Determinants of human longevity: parental age at reproduction and offspring longevity. Longevity Rep10(54):7–14

Jacquard A (1982) Heritability of human longevity. In: Preston SV (ed) (Biological and social aspects of mortality and the length of life. Liège, Ordinia Edition, pp 303–313

Kannisto V (1991) Frailty and survival. Genus 47(3–4):101–118

Le Bourg E, Thon B, Légaré J, Desjardins B, Charbonneau H (1993) Reproductive life of French-Canadians in the 17–18th centuries: a search for a trade-off between early fecundity and longevity. Exp Gerontol 28:217–232

McGue M, Vaupel JW, Holm N, Harvald B (1993) Longevity is moderately heritable in a sample of Danish twins born 1870–1880. J. Gerontol 48(6):B237–B244

Murphy EA (1978) Genetics of longevity in man. In: Schneider E (ed) Genetics of Aging. New York, Plenum Press, pp 291–301

Pearl R (1931) Studies on human longevity. IV. The inheritance of longevity. Preliminary report. Human Biol 3:245–269

Pearl R, Pearl Dewitt R (1934) Studies on human longevity. VI. The distribution and correlation of variation in the total immediate ancestral longevity of nonagerians and centenarians, in relation to the inheritance factor in duration of life. Human Biol 6:98–222

Philippe P (1980) Longevity: some familial correlates. Social Biol 27(3):211–219

Registre des Naissances (1893) Arles, IIE 318

Registre des Naissances (1894) Arles, IIE 321

Registre des Naissances (1895) Arles, IIE 323

Robine JM, Allard M (1995) Validation of the exceptional longevity case of a 120 year old woman. Facts Res Gerontol 363–367

Genetics of Aging

F. Schächter[*]

Abstract

A framework is proposed to interpret the findings from a growing number of genetic studies of human longevity. The comparative analysis of the genetic structure of long-lived populations provides indications about the genetic variants that affect survival. Such results stemming from population genetics approaches are significant on a statistical level but point to various mechanisms of aging that are active in certain individuals. Criteria are needed to relate the differential genetic effects on survival and individual aging. We propose a general definition of aging, based on its common denominator, that is valid for the organism and the cell. We argue that aging is discontinuous and occurs in discrete steps. In search of a "primum movens" for aging, the concept of compensatory adaptation is introduced, which implies a new classification of aging processes. From an evolutionary perspective compensatory adaptation reflects a lack of natural selection. Ambivalent aspects of a longevity determinant and an aging target are tied in the genome.

Introduction

Every gerontologist knows that aging is "very complex" and that aging processes affect all levels of biological organization of a living organism and display an endless variety among species. In this jungle of phenomena, it has proven difficult to identify causes and separate them from consequences. Most of the available data on aging are descriptive. Therefore, it seems worthwhile trying to find criteria to classify aging processes, if not in a perfect causal order, at least on a causal gradient. Much confusion has been generated in this field by the undefined use of the words "aging" or "senescence" in different contexts. It is a fundamental prerequisite to clarify the various meanings of "aging" and reflect on their relationships.

Pre-Socratic wisdom whispers to us "All men follow Gompertz survival curves, you are a man, therefore you must slide down a Gompertz slope." Some-

[*] CEPH, 27 rue Juliette Dodu, 75010 Paris

J.-M. Robine et al. (Eds.)
Longevity: To the Limits and Beyond
© Springer-Verlag Berlin Heidelberg New York 1997

thing sounds wrong in this syllogism, though. Death is the end-product of aging. The formal definition of aging as "an increase in the probability of dying" has one major drawback: it is a probabilistic definition that relies implicitly on statistical observations and is based on a future event! If one adds to that the idea that death is eventually caused by aging, the circularity of this definition becomes obvious. One would like direct measures of aging rather than a propensity dependent upon statistical validation. A theoretical framework for such measures is proposed here.

Aging takes place within a certain time-frame that appears to be determined by the genetic make-up of the species. Evolutionary theories of aging provide a rationale for differences in life span, as an element of life-history strategies, among species. The Chronos Project aims at identifying in humans the genetic components of inter-individual differences in life span. The new challenge will be to interpret these genetic influences: how do they affect survival?

Extreme Longevity

The Chronos Project and a growing number of similar undertakings are searching for genetic factors in long-lived human groups. Two main forces underlie these investigations. Firstly, the power of detecting genetic influences depends on the extent of selection by survival. According to estimates in the 1980s, about 2 % of a cohort of sexagenarians would reach the age of 100: the genetic structure of a population of centenarians is the outcome of passing through this sieve. Secondly, one hopes to gain insight into normal aging rather than merely age-related pathologies, because exceptional survivorship should involve not only escape from major disease but also a reduced rate of aging.

In our first association study in centenarians, we found distortions in the allelic and genotypic distributions at the apolipoprotein E (APOE) and angiotensin-converting enzyme (ACE) loci (Schächter et al. 1994). Both of these gene products have pleiotropic effects in the organism. It is interesting to note that apolipoprotein E, involved in the re-distribution of cholesterol among cells within and between tissues, may be considered a maintenance protein (Mahley 1988). Its variants have an impact on susceptibility to cardiovascular disease (Davignon et al. 1988) and Alzheimer's disease (Saunders et al. 1994). Our results seemed compatible with the notion that ε4 is a risk factor and ε2 a protective factor in these two spheres of age-related pathologies. This interpretation however may hide a more complex reality. Indeed, it appears from genotyping different age groups that the influence of the ε4 variant on survival may change its sign, not just once but three or four times (Poirier et al. 1993)! The negative effect on survival in the seventh decade of life would become positive in the eigth decade, then negative again in the ninth decade; recent results from the Danish super-semicentenarians, aged between 105 and 110, suggest that this effect may change once more (Bernard Jeune, personal communication)! Although cross-sectional studies are always susceptible to biases and the number of subjects in each age group is not very big, this finding is in keeping with the changing impact of cholesterol levels on survival. Furthermore, the main message is clear: relative allele-specific effects on survival are age-dependent. The distributions found in centenarians are the net result of such selections over a few decades. Because of the increase in mortality rates, the contribution of these selective differences increases with age so that even small relative effects in the ninth decade are likely to prevail in the final observed distribution. Concerning ACE, our results were unexpected since we found the genotype ACE/DD[1], previously identified as conferring a risk for myocardial infarction in an otherwise low risk population (Cambien et al. 1992) – a group to which potential centenarians would belong – to be more represented in centenarians. This result implies a change, somewhere between 50 and 100 years of age, in the survival value of a variant associated with the ACE/D polymorphism. Worthy of mentioning, other recent studies provide evidence for the view that genetic factors classically deemed "unfavorable" for the cardiovascular system may promote longevity (Mari et al. 1995 and Daniela Mari, personal communication).

The idea that the genome is loaded with "late-acting deleterious variants" that have accumulated for lack of selective pressure has been around for a long time. The emerging picture that may be anticipated from these first ventures on the terra incognita of extreme longevity would instead speak of unequal and age-dependent survival values of certain variants. How does that relate to aging at the individual level?

Individual Aging

Survival curves are averaged for many heterogeneous factors: genetic background, environmental conditions, history. Would a clone of genetically identical individuals follow a Gompertz survival curve? Such a question pertains to mechanisms of aging at the individual level. The statistical patterns observed in populations, from which mortality rates and Gompertz parameters are derived, should be inferred from a knowledge of individual aging, in the same fashion as classical thermodynamics has been deduced a posteriori from molecular physics. This problem is of long-standing interest to biologists and epidemiologists; the concepts of "frailty" or "liability" have been invented for that reason. Various overlapping definitions of aging exist, such as "decrease in physiological functions and homeostasis" or "decrease in adaptive capacity and response to stress," etc. Although providing an intuition of the aging phenomenon, these definitions are too imprecise and do not lend themselves to formal developments.

We can conceive of an organism as a complex biological system with the representation, well-known in physics, of a phase space. This is a multidimensional

---

[1] D is an allele of the ACE gene corresponding to a deletion in the gene, it can be visualised by amplifying a segment of DNA in this region, yielding a shorter length.

space in which each point corresponds to a unique state of the system, formally described by the positions and speeds of all the particles that make up the system. When an organism is stressed, it means that it is pushed toward some border of this phase space and then has to return somewhere at a more comfortable point. During aging, the space available for exploration and the paths from one point to another dwindle. This is a primary observation common to all phenomena of aging. If the level of environmental risk is constant, that is, the frequency, magnitude and nature of perturbations, this shrinking of the phase space will result in increased mortality, simply because the system will be pushed more readily out of the states compatible with continued living. In fact, an aging organism will adapt its behavior and environment in order to reduce the risk, and that is also a primary observation in aging, of paramount importance for understanding the pattern of age-related changes. It has been said about Jeanne Calment that she is on the way to becoming petrified: she does not move, has nearly stopped eating, her metabolism is close to zero – and so she may reach immortality!

So this shrinkage of the phase space results in increased vulnerability and that is aging, whatever form it takes. How could it be measured? This is the purpose of biomarkers. An ideal biomarker should reflect the increase in vulnerability. There are grounds for thinking that individual vulnerability is not a smooth function of time, but rather a discrete function that undergoes sharp transitions. Extreme examples may be taken from species that undergo "rapid senescence and sudden death," like the Pacific salmon, with a catastrophic post-reproductive senescence (Finch 1990). In humans, many age-related pathologies have an age of onset that is under stringent control. Several related evolutionary theories of aging – the mutation accumulation, the antagonistic pleiotropy, the disposable soma theories – tend to postulate that the genome is loaded with "late-acting deleterious variants." Indeed, the recent and massive increase in life expectancy in developed countries has unmasked a plethora of age-related pathologies. But what really lies beneath the concept of "late-acting deleterious variant?"

Let us consider a gene that has several variants common in the general population. In most cases, they will have equal fitness values; otherwise the distribution would not be in Hardy-Weinberg equilibrium and would still be evolving under selection pressure. The context of this gene undergoes changes during aging, until one day the variants cease to behave equally with regard to survival: a differentiation appears in the relative survival values of hitherto equivalent alleles. When a change in boundary conditions in the organism – be it blood pressure, serum levels of glutathione or of tumor necrosis factor, etc. – affects gene expression in a sufficiently coherent fashion in many cells, one may expect such splitting of genetic variants' survival value to occur in those genes that interact with the altered boundary conditions.

The discrete character of age-related changes is even more obvious at the cellular level. What appears in a population as a progressive loss of proliferation and has been termed cellular senescence (wrongly according to some gerontologists) may reflect shifts in the distribution of differentiation states (Macieira-Coelho 1995a). What is a differentiation state? It may be defined as a characteristic pattern of gene expression. In the phase space of a cell, differentiation states represent a tiny subset of points accessible for this specific cell, certainly not a continuum. For a dividing cell, the discrete jumps are transitions from one state of differentiation to the next. This has been interpreted by Remacle, within the theoretical frame of irreversible thermodynamics, as reorganization into states of lower entropy production (Toussaint et al. 1991). The cell that reaches the end of its proliferative life span and goes into irreversible Go is buying time for survival. Indeed, if these cells are forced into division, they undergo massive death. Alvaro Macieira-Coelho has shown that the last mitosis preceding the irreversible arrest of division in human fibroblasts is quite catastrophic (Macieira-Coelho 1995b). During this mitosis – suggestively called the "quantal mitosis" – the distribution of DNA between daughter cells is perturbed and the chromatin changes its organization abruptly. Subsequent enforced division by viral transforming proteins, like the SV40 large T antigen, results in cell death and, for the few survivors that pass the barrier to immortalization, chromosomal instability and aneuploidy. In contrast, aging for a post-mitotic cell is the more subtle and progressive accumulation of damage leading beyond a certain threshold to cell death.

At the organism level, vulnerability increases with age in discrete steps, in a pattern that is specific for each individual and depends on genetic, environmental and historical factors. It may be useful to split this increase in vulnerability into two components: the discrete steps, due to environmental perturbations and the onset of deleterious gene action in any combination, and a continuous underlying accumulation of damage that would be the common denominator, "intrinsic aging."

Compensatory Adaptation

Let us now magnify the response of a biologic system to perturbation. The generic profile would be an increase in vulnerability, followed by a return to or near the initial state. That a perturbation always entails an immediate increase in vulnerability is easily seen if we conceive of a complete defect in homeostatis: the system could not endure any perturbation. The spring-back return is the homeostatic response. There are three possibilities: 1) the phase space remains identical – this is perfect homeostasis, there is no change in vulnerability; 2) the phase space shrinks – this is imperfect homeostasis, or hysteresis and goes with an increase in vulnerability; and 3) the phase space enlarges – this would be training, or positive adaptation entailing reduced vulnerability. I argue that the homeostatic response generally entails an increase in vulnerability in organisms beyond a certain age, depending on the initial level of vulnerability and the magnitude of the stress. If the homeostatic capacity of the system is still good, it will adapt to a chronic stress, to the extent of canceling its effect. This is what I call compensatory adaptation, and this brings about a reorganization of the system into a state of higher vulnerability. A majority of commonly observed age-related changes may fall into this category. They are so pervasive at all levels of biological organization that they often hide the primary causes of aging while creating the familiar phenotypes. For example, at the cellular level, inhibitors of cell division found in senescent cells may not result from a "programmed senescence" but rather may be induced when the cells are unable to undergo further mitosis without genomic damage. At the tissue level, enlargement of cardiomyocytes may be stimulated by stress-induced growth factors to compensate for cell loss. At the organism level, higher release of pro-inflammatory cytokines or higher blood pressure is the consequence of many steps of compensatory adaptation. When the vulnerability is too high for adaptation to occur, a vicious circle sets in with rapid aging, as in the development of a fatal disease.

This general picture should not ignore the fact that compensatory adaptation may bring about some intriguing pro-survival reorganizations. A striking example of this is provided by Brattleboro rats (Van Leeuwen et al. 1989). These rats are homozygous for a recessive mutation in the vasopressin (VP) gene, normally expressed in specialized neurons, that results in the inability of these neurons to synthesize active vasopressin and causes diabetes insipidus. Surprisingly, some revertants arise during aging. The molecular mechanism involves a recombination between the neighboring highly homologous oxytocin and vasopressin genes, thereby creating heterozygous neurons that are able to express vasopressin, and the number of these revertant neurons increases with aging. Since the revertant neurons are under chronic osmotic stress to express VP, the gene in these cells is probably transcriptionally hyperactive while turning out an inactive protein, and this state of hyperactive transcription may promote recombination with the oxytocin gene. In that case, the compensatory adaptation enlarges the phase space of the cells and the organism in which it occurs, reduces vulnerability and increases survival – but importantly, the initial point was a genetically defective, sick system. Indeed, when compensatory adaptation takes place in a "fit" organism, it is most likely to bring it downhill.

Natural Selection and Aging

Let us come back to the phase space of an organism, adding the time dimension, and let us consider the complete phenotype in one block, unfolded in space and time. The pre-reproductive section of the phenotype has been perfected by natural selection, whereas the post-reproductive part of the phenotype has been neglected. A hallmark of natural selection optimization is a good homeostatic capacity, an adequate response to challenges, giving the impression that every situation has been predicted, that all the components of the organism are useful, well coordinated and under control. This is the domain of harmony between structure and function. This is what Sir Karl Popper has termed "implicit knowledge": the adequate reaction to random exterior circumstances. That beautiful organization, crystallized under the pressure of natural selection, melts away in the aging portion of the phenotype, to be replaced by compensatory adaptation, one might say by self-organization. In that sense, aging is the self-organization of the soma left to its own means, unaided by natural selection. Meanwhile, ceaseless endogenous processes are at work, steadily increasing the vulnerability, among which the most intrinsic may be the relentless attacks on the genome.

At the Core of Cause and Effect: The Genome

When searching for variants that have differential effects on survival, one is looking at the genome from the viewpoint of a determinant of longevity. This idea is well rooted, that somehow longevity is "determined," in either an active or passive fashion, by the genome: species maximum life span and individual longevity. But the genome also happens to be a major target of aging processes, indeed of primary aging processes directly deriving from intrinsic cumulative damage (Slagboom 1990).

On the one hand, recombination and mutation during meiosis generate the diversity on which natural selection will operate. On the other hand, too much recombination would destroy the winning combinations! And mutations may create weaker phenotypes and disease. Striking the right balance is the topic of a classical problem in evolutionary biology: what is the optimal rate of mutation? One may ask, in a wider perspective, what is the optimal genomic plasticity? Common mechanisms may underlie genomic plasticity during meiosis and during mitosis; therefore the observable consequences may be tied by common constraints.

For a species living in an adverse environment with high risk, the preferred life strategy may be to invest heavily in reproduction and evolve rapidly. This implies consequences at the genomic level, not merely as a byproduct of selection, but because overlapping mechanisms may determine the rate of evolution and the rate of aging.

References

Cambien F, Poirier J, Lecerf L, Evans A, Cambou JP, Arveiler D, Luc G, Bard JM, Bara L, Ricard S, Tiret L, Amouyel P, Alhenc-Gelas F, Soubrier F (1992) Deletion polymorphism in the gene for angiotensin-converting enzyme in a potent risk for myocardial infarction. Nature 359:641–644

Davignon J, Gregg RE, Sing CF (1988) Apolipoprotein E polymorphism and atherosclerosis. Arteriosclerosis 8:1–21

Finch CE (1990) Longevity, senescence and the genome. The University of Chicago Press, London, pp 43–120

Macieira-Coelho A (1995a) The implications of the "Hayflick limit" for aging of the organism have been misunderstood by many gerontologists. Gerontology 41:94–97

Macieira-Coelho A (1995b) Chaos in DNA partition during the last mitoses of the proliferative lifespan of human fibroblasts. FEBS Letters 358:126–128

Mahley RW (1988) Apolipoprotein E: Cholesterol transport protein with expanding role in cell biology. Science 240:622–630

Mari D, Mannucci PM, Coppola R, Bottasso B, Venturati M, Bauer KA, Rosenberg RD (1995) Hypercoagulability in centenarians: the paradox of successful aging. Blood 85:3144–3149

Poirier J, Davignon J, Bouthillier D, Kogan S, Bertrand P, Gauthier S (1993) Apolipoprotein E poly-morphism and Alzheimer's disease. Lancet 342:697–699

Saunders AM, Schmader K, Breitner JCS, Benson MD, Brown WT, Goldfarb L, Goldbager D, Man-waring MG, Szymanski MH, McCown N, Dole KC, Schmechel DE, Strittmatter WJ, Pericak-Vance MA, Roses AD (1994) Apolipoprotein E e4 allele distributions in late-onset Alzheimer's dis-ease and in other amyloid-forming diseases. Lancet 342:710–711

Schächter F, Faure-Delanef F, Guénot F, Rouger H, Froguel P, Lesueur-Ginot L, Cohen D (1994) Genetic associations with longevity at the APOE and ACE loci. Nat Genet 6:29–32–29–30

Slagboom PE (1990) The aging genome: determinant or target? Mut Res 237:183–187

Toussaint O, Raes M, Remacle J (1991) Aging as a multi-step process characterized by a lowering of entropy production leading the cell to a sequence of defined stages. Mech Ageing Devel 61:45–64

Van Leeuwen F, Van der Beek E, Seger M, Burbach P, Ivell R (1989) Age-related development of a hete-rozygous phenotype in solitary neurons of the homozygous Brattleboro rat. Proc Natl Acad Sci USA 86:6417–6420

Oxidative Stress May Be a Causal Factor in Senescence of Animals

*R.J. Mockett and R. S. Sohal**

Abstract

The oxidative stress hypothesis attributes age-related increases in mortality rates to physiological attrition, which is underlaid, in turn, by oxidative molecular damage. This damage results from an imbalance between prooxidant production, antioxidant defenses and repair processes. The hypothesis is supported by age-related increases in oxidative stress and damage to various biological macro-molecules. There is an inverse relationship between oxidative stress and life span potential, both within and among species. Tandem overexpression of antioxidant enzymes increases median and maximum life spans, and decreases the levels of oxidative stress and molecular damage. Prooxidant production appears to be most closely linked with rates of aging, suggesting that the most effective interventions will minimize the levels of prooxidant production rather than bolstering antioxidant defenses alone.

Introduction

Aging is a process of unreversed deteriorative change in living systems, which results in an increasing probability of death as a function of time. Many theories have been advanced purporting to explain the cause of this process, including tissue- or species-specific, programmatic, random damage-based, and evolutionary theories (Finch 1990). For a number of reasons, the oxidative stress hypothesis, which is the subject of this review, is currently eliciting considerable interest (Ames et al. 1993; Stadtman et al. 1992; Sohal and Orr 1995). According to this hypothesis, aging is the result of a chronic imbalance between prooxidant production, antioxidant defenses and repair processes. Prooxidants, primarily reactive oxygen species (ROS) produced as by-products of mitochondrial respiration, inflict damage on biological macromolecules unless they are first inactivated by enzymatic or small molecular weight antioxidants. Some of this damage is repaired, or the damaged molecules are degraded, but the remaining damage accumulates and causes physiological attrition. If the imbalance increases, due to

* Department of Biological Sciences, Southern Methodist University, Dallas, TX 75275

J.-M. Robine et al. (Eds.)
Longevity: To the Limits and Beyond
© Springer-Verlag Berlin Heidelberg New York 1997

enhanced prooxidant production or impairment of defense or repair processes, then the physiological attrition will accelerate. Even if the imbalance is constant, i.e., there is no change in prooxidant production, antioxidant defenses or repair processes, the constant accumulation of damaged molecules will still produce a linear increase in physiological attrition and this hypothesis may still identify the primary causal factor in aging.

The attractive features of the oxidative stress hypothesis are manifold. For example, the proposed mechanism is sufficiently general to account for the existence of senescence in a broad range of phylogenetic groups. It is sufficiently specific that experimental tests can be conducted to validate or falsify it. Consequently, and most importantly, because many such tests have already been conducted, substantial supporting evidence already exists, so the hypothesis is not arbitrary. The following predictions of the oxidative stress hypothesis have been systematically tested in this laboratory:

1) The level of oxidative stress increases as a function age within multiple species of different phyla, as evidenced by increasing ROS production or decreasing antioxidant defenses/repair processes;
2) The levels of oxidatively damaged lipids, proteins and nucleic acids increase during aging;
3) Differences in life spans between species are associated with different rates of prooxidant production or antioxidant defense capabilities;
4) Conditions that alter the life span of a given species do so by lessening the degree of oxidative stress, or are at least associated with decreased oxidative stress and oxidative molecular damage.

Oxidative Stress and Aging

Mitochondria are believed to be the main site of ROS production, since they account for approximately 90 % of intracellular oxygen consumption and convert 2–3 % of this oxygen to $O^{\cdot}_2$ (Chance et al. 1979). The product of $O^{\cdot}_2$ dismutation by mitochondrial Mn superoxide dismutase (SOD) is hydrogen peroxide, H_2O_2. This molecule diffuses across the mitochondrial membranes and undergoes a Haber-Weiss or Fenton reaction, generating the hydroxyl free radical, $\cdot OH$. This radical is believed to interact directly with protein, lipid or nucleic acid molecules, initiating a destructive free radical chain reaction.

Since $\cdot OH$ is too reactive to accumulate at detectable levels, shifts in the level of oxidative stress are estimated based on measurements of other ROS and antioxidant defenses, and products of the reactions of ROS with biological molecules. In submitochondrial particles isolated from rats and mice of various ages, the rate of $O^{\cdot}_2$ production was found to be higher in the liver, heart, brain and kidney of the older animals (Sohal et al. 1990b, 1994; Fig. 1 A). H_2O_2 release by heart and brain mitochondria was elevated 21 % and 30 %, respectively. In houseflies, the rate of mitochondrial H_2O_2 production increased exponentially between the time

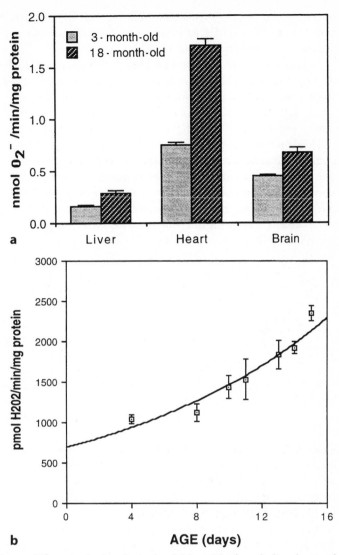

Fig. 1 A. Comparison of rates of O_2^- generation by submitochondrial particles from the liver, heart and brain of 3- and 18-month-old rats, measured as SOD-inhibitable reduction of ferricytochrome c. (From Sohal et al. 1990b) B. Rates of H_2O_2 release by housefly mitochondria, measured as oxidation of p-hydroxyphenylacetate during the enzymatic reduction of H_2O_2 by horseradish peroxidase. (From Sohal and Sohal 1991) Values are mean ± SEM of three to six measurements

of maturation (4 d) and the onset of the rapid dying phase (16 d). The absolute level of H_2O_2 generation increased 100 % in this interval (Sohal and Sohal 1991; Fig. 1 B). Mitochondrial H_2O_2 production showed a similar age-related increase in the fruit fly, *Drosophila melanogaster* (Sohal et al. 1995 a). Exposure of housefly mitochondria to experimental oxidative stress, in the form of 100 % O_2 or x-irradiation, produced a comparable increase in H_2O_2 production (Sohal and Dubey 1994). Thus, ROS production appears to increase as a function of age in different species from at least two different phyla.

Antioxidants are the second component of the equation determining the level of oxidative stress. Small molecular weight antioxidants include glutathione, tocopherols and ascorbate. Enzymatic antioxidants include MnSOD, cytosolic Cu-Zn SOD, catalase and glutathione peroxidase. The antioxidants prevent or terminate some of the free radical chain reactions resulting from ROS production.

Measurements of various antioxidants in different species and as a function of age have shown conflicting trends (Sohal and Orr 1992; Sohal et al. 1990c). The total level of antioxidant defense has been difficult to assess because there are so many components, many of them compartmentalized, and the relative importance of each component is likely to differ between tissues. To circumvent this problem, the ratios of redox couples, such as GSH/GSSG, NADH/NAD$^+$ and NADPH/NADP$^+$, were measured. The ratios became more prooxidizing as a function of age in insects and mammals (Sohal et al. 1990a; Noy et al. 1985), suggesting a decreased defensive capability in response to the increased ROS production. This conclusion was reinforced by exposing houseflies to experimental oxidative stress, in the form of x-irradiation. The older flies experienced greater losses of enzyme activity and greater increases in oxidative damage in response to a given x-ray dosage, implying that their defenses were impaired relative to those of the younger flies (Agarwal and Sohal 1993).

The final component of the equation is the capacity to repair damage inflicted by ROS. Although most repair processes have not been systematically studied as a function of age, there is evidence for a decrease in the activities of alkaline proteases (Agarwal and Sohal 1994a; Starke-Reed and Oliver 1989). Induction of protease activity by acute oxidative stress (100% oxygen) is also lost during aging. Thus, the combined evidence from studies on ROS production, antioxidant defenses and repair processes suggests that oxidative stress increases during aging.

Oxidative Damage and Aging

If oxidative stress is a crucial causal factor in aging, then levels of oxidative damage to lipids, proteins and DNA should increase with age. Since the level of oxidative stress appears to increase with age, the corresponding rate of accumulation of damage should also increase, i.e., the total level of damage should increase exponentially.

Exhalation of alkanes, such as n-pentane and ethane, is a reliable indicator of the rate of peroxidation of polyunsaturated fatty acids in biological membranes. The rate of n-pentane exhalation increases approximately 50% during adult life of the male housefly (Sohal et al. 1985) and is also increased in tissues of the rat. There is also some evidence that lipofuscin, the autofluorescent material that accumulates in most cell types during aging, is a product of oxidative damage of lipids and proteins (Sohal et al. 1989a).

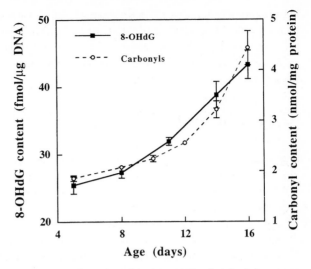

Fig. 2. Accumulation of 8-OHdG in total DNA and protein carbonyl content in whole body homogenates of houseflies. DNA damage, determined by HPLC, and protein damage, determined spectrophotometrically following exposure to 2,4-dinitrophenylhydrazine, are correlated with each other and show exponential increases during aging. (From Agarwal and Sohal 1994b)
ipids and proteins (Sohal et al. 1989a).

Protein oxidative damage is most commonly measured as the level of carbonyl content. Carbonyl content increases as a function of age in *Drosophila* (Orr and Sohal 1994) and in gerbil brain, heart and testis (Sohal et al. 1995b). It increases exponentially in whole-body homogenates and flight muscle mitochondria of male houseflies (Sohal and Dubey 1994; Sohal et al. 1993a; Fig. 2). Carbonyl content increases with age in specific regions of the mouse brain, where it is associated with losses of cognitive and motor skills (Dubey et al. 1996), and in human brain (Smith et al. 1991). The fraction of total protein inactivated by carbonyl formation has been estimated to rise from 10% to 30% in rat hepatocytes during aging of the animal (Starke-Reed and Oliver 1989; Stadtman 1992).

This finding is consistent with the observations that inactive proteins accumulate during aging, that activities of various enzymes decrease during aging, and that metal-catalyzed oxidations involving ROS and transition metal cations inactivate several of these enzymes *in vitro* (Fucci et al. 1983).

DNA oxidative damage during aging has been measured in various ways, with single strand breaks, single-stranded regions and oxidized bases increasing in multiple organs of several mammalian species (Holmes et al. 1992). The accumulation of 8-hydroxydeoxyguanosine (8-OHdG), a product of ROS reactions with guanine residues, follows the same age-related exponential increase as protein carbonyl content in male housefly homogenates (Agarwal and Sohal 1994b; Fig. 2). Experimental oxidative stress, in the form of x-irradiation or exposure to 100% O_2, causes similar increases in both carbonyl and 8-OHdG content (Sohal and Dubey 1994; Agarwal and Sohal 1994b). Both nuclear and mitochondrial

DNA oxidative damage increase during aging in *Musca*, and 8-OHdG content increases steeply during aging of *Drosophila* (Sohal et al. 1995a). In summary, these results show that the age-related increase in oxidative stress is associated with increasing levels of oxidative molecular damage in insects and mammalian species.

Life Span Variations Among Species

If oxidative stress is a general causal factor in aging, then variations in levels of oxidative stress and damage should be associated with differences in the duration of life in different species. Changes in maximum life span potential (MLSP), rather than average life span, are most salient in establishing a change in the rate of aging. Changes in average life spans may reflect optimization of living conditions or removal of a specific cause of premature death, rather than aging *per se*. This point is adequately demonstrated by the doubling of the average human life expectancy in developed countries in the last 200 years. This increase was produced entirely by the prevention or cure of specific disease processes, with no change in either the underlying rate of aging or the maximum life span (Fries 1980).

The rate of living theory (Pearl 1928) predicts an inverse relationship between the length of life and the rate of energy metabolism. This theory has been validated in many mammalian species (Sacher 1977). Assuming roughly constant rates of ROS generation during production of a given amount of energy in different species, the rate of living theory is consistent with the oxidative stress hypothesis. Indeed, the oxidative stress hypothesis provides a mechanistic explanation for the inverse relationship between rate of living and MLSP. Departures from the assumption of constant ROS generation during metabolism, e.g., due to structural differences in inner mitochondrial membrane components, could account for the observed deviations from the rate of living generalization.

This idea was tested by measuring ROS production and antioxidant defenses in young adult animals of different species. A hyperbolic relationship was observed in rates of $O_2^{\cdot}$ production in submitochondrial particles (smps) from the housefly (MLSP = 0.25 yr) and five mammalian species with MLSPs ranging from 3.5 to 30 years (Sohal and Orr 1995; Sohal et al. 1989b; Fig.3 A). A similar relationship was observed for mitochondrial H_2O_2 production in dipteran insects and in the kidney and heart of six mammalian species (Sohal et al. 1990c, 1995c; Fig. 3 B). There were also direct relationships between the rates of mitochondrial $O_2^{\cdot}$ generation and state 4 respiration, and between specific metabolic rate (SMR; cal/g/day) and respiration rate (Ku et al. 1993).

The relationship between antioxidant defenses and MLSP was more ambiguous, as were the results for antioxidant levels as a function of age (Sohal et al. 1990d). SOD and catalase activities in the liver, heart and brain were positively correlated with MLSP, but glutathione showed a negative correlation and glutathione peroxidase had a tissue-specific relationship. The overall relationship for

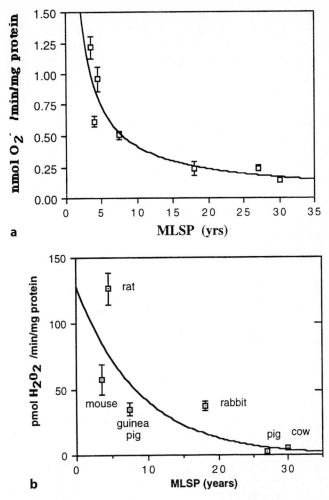

Fig. 3 A. Mitochondrial O_2^- generation in the kidney of seven mammalian species shows a negative correlation with maximum life span potential (MLSP). (From Ku et al. 1993) B. The rate of H_2O_2 release by liver mitochondria shows a similar relationship with MLSP. (From Sohal et al. 1990d) Values are mean ± range or SEM of two to nine determinations. MLSPs, in years, are: mouse, 3.5; hamster, 4; rat, 4.5; guinea pig, 7.5; rabbit, 18; pig, 27; cow, 30

antioxidants, if any, was difficult to establish, but the striking trends for ROS production imply a negative correlation between oxidative stress and MLSP.

The relationships between oxidative stress, molecular damage and MLSP were reinforced by comparisons between two additional pairs of species with pronounced differences in MLSP and metabolic potential (amount of energy consumed per g body weight during adult life). The white-footed mouse, *Peromyscus leucopus*, has a 2.3-fold greater MSLP than the house mouse, *Mus musculus*, accompanied by a higher SMR. *Peromyscus* mitochondrial O_2^- production, H_2O_2

release and protein carbonyl content were significantly lower, while SOD, catalase and glutathione peroxidase activities were higher than those of *Mus* (Sohal et al. 1993b). Similar results were observed in the rat and pigeon, which have the same SMR but seven to eight-fold differences in metabolic potential and MLSP (Ku and Sohal 1993). The only deviations from the trends reported in *Peromyscus* and *Mus* were lower levels of catalase in the longer-lived pigeon and identical levels of protein carbonyl content in the rat and the pigeon. Notwithstanding these results, the rates of ROS production and other antioxidant defense levels showed the predicted relationship with MLSP in the rat and pigeon.

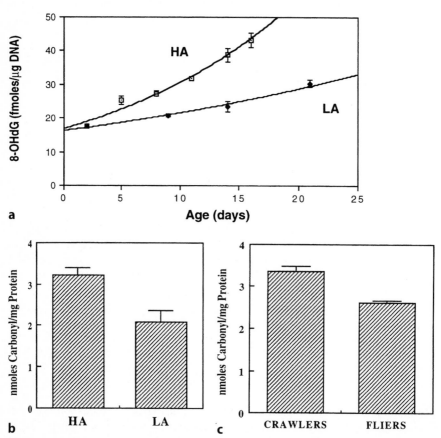

Fig. 4A. Total DNA oxidative damage (8-OHdG) increases more quickly in high activity (HA) flies than in low activity (LA) flies. In each case, damage accumulates as an exponential function of chronological age. Values are mean ± SEM of 3–6 determinations. (From Agarwal and Sohal 1994b). B. Protein carbonyl content is higher in homogenates of 14-day-old houseflies kept under HA conditions than in those kept under LA conditions. C. Protein carbonyl content is also higher in homogenates of 12-day-old crawlers than in fliers of the same age. Values are mean ± SEM of four determinations. (From Sohal et al. 1993a)

Fig. 5. Mitochondrial H_2O_2 release is lower in the brain, heart and kidney of mice on dietary restriction (DR) than those fed *ad libitum* (AL). Asterisks indicate $P < 0.001$ for differences between AL and DR regimes and for age-related differences. The release of H_2O_2 increases with age in all tissues studied, and is higher in the AL than the DR group at 9, 17, and 23 months of age. Results are mean ± SEM of four to six measurements. (From Sohal et al. 1994)

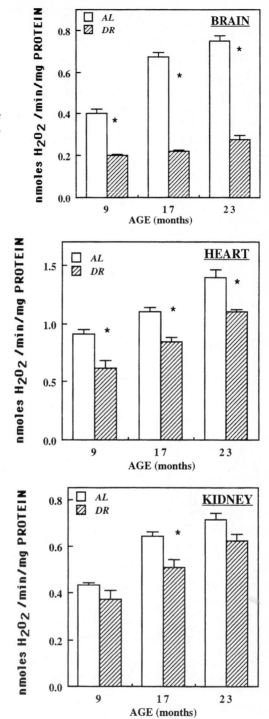

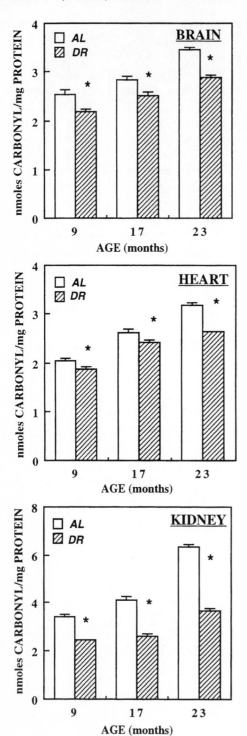

Fig. 6. The protein carbonyl content of brain, heart and kidney homogenates is higher in AL than in DR mice at 9, 17 and 23 months of age (asterisks indicate P < 0.001). The discrepancy in carbonyl content between AL and DR mice also increases with age. An exponential age-related increase in carbonyl content is apparent, particularly in the kidney and in the brain of AL mice. Results are mean ± SEM of three to ten determinations. (From Sohal et al. 1994)

Life Span Variations Within Species

A variety of conditions and experimental regimens result in differences in life span within a cohort population. For instance, changes in ambient temperature alter the metabolic rate in poikilotherms, prolonging the life span of insects 2- to 3-fold without affecting the metabolic potential (Sohal and Orr 1995). Since the rate of aging determines the life span, changes in the metabolic rate, which is directly proportional to rates of ROS generation in different species, presumably govern the rate of aging.

Comparisons can also be made between individuals of the same chronological age but different physiological ages (i.e., proportion of the life span completed). In houseflies, both the average and the maximum life spans can be doubled by prevention of flight activity. Flies kept under low activity conditions have lower rates of O_2^- production and H_2O_2 release, lower levels of protein carbonyl content, and lower levels of 8-OHdG, in both nuclear and mitochondrial DNA, than flies of the same chronological age kept under conditions of high activity (Sohal et al. 1993a; Agarwal and Sohal 1994b; Fig. 4A, B).

The observation that all flies lose the ability to fly a few days prior to death is the basis of a second strategy for separating chronological and physiological ages. Thus, the "crawlers" in a cohort have completed a much greater fraction of their life span than the "fliers". Mitochondrial H_2O_2 release was found to be twice as high in crawlers as in fliers of a given chronological age, while protein carbonyl content was significantly lower in the fliers (Sohal et al. 1993a; Fig. 4C).

Dietary restriction is the only experimental regimen known to produce a significant and reproducible increase in MLSP in mammals. In comparison with *ad libitum*-fed (AL) rodents, those on dietary restriction (DR) consume 40% fewer calories and live 35–40% longer (Sohal et al. 1994; Richardson 1991; Weindruch and Walford 1988). DR mice have a lower rate of respiration than AL mice, accompanied by lower rates of mitochondrial O_2^- and H_2O_2 production (Fig.5) and lower levels of protein carbonyl content (Fig. 6). These relationships are observed in multiple organs at several different chronological ages (9, 17 and 23 months; Sohal et al. 1994).

Experimental Increases in Antioxidant Defenses

Overexpression of antioxidant enzymes is a final mechanism by which MLSP can be altered within a species. Unlike any of the other conditions described, this mechanism involves the creation of transgenic animals, which modifies the genotype as well as the environment. An important consequence is that, whereas the other regimens raise MLSPs by lowering the metabolic rate, this regimen raises MLSP by raising both the metabolic rate and the metabolic potential (Sohal et al. 1995a). Thus, the increased life span is potentially more beneficial to the animal and provides more direct support for the oxidative stress hypothesis.

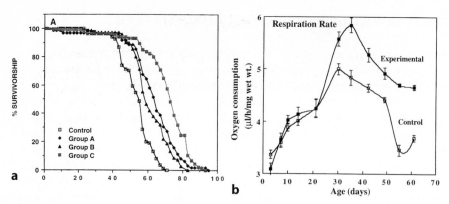

Fig. 7A. Survivorship curves for three groups of adult male transgenic *Drosophila melanogaster* over-expressing single extra copies of Cu-Zn SOD and catalase genes (groups A, B, and C). The control group has only the normal diploid complement of antioxidant genes. (From Orr and Sohal 1994) B. The *in vivo* rate of oxygen consumption by transgenic strain "A" is greater than that of control flies. In conjunction with their longer life spans, the increased rate of oxygen consumption by transgenic flies results in a 30% difference in metabolic potential. Values are mean ± SE of 4–28 determinations. (From Sohal et al. 1995a)

Initial efforts to increase antioxidant defenses focused on dietary supplementation with small molecular weight antioxidants. These experiments did not show a consistent increase in MLSP, which suggested that endogenous antioxidant levels provided sufficient defenses against oxidative stress *in vivo*. A second interpretation, consistent with findings that ROS production was more closely associated with aging and inter-species variations in MLSP, was that ROS production is the more crucial parameter determining levels of oxidative stress and damage, and the rate of aging. Thus, if reactive ·OH is only formed at sites where it immediately inflicts irreversible damage, then increasing levels of antioxidants would have a minimal effect on life span. Alternatively, the level of antioxidant defenses might be regulated by the cell, so that increasing the exogenous dose depressed the levels of endogenous defenses.

Initial experiments overexpressing single antioxidant enzymes (Cu-Zn SOD or catalase) yielded similar results. The effects on life span were small, reinforcing the idea that ROS production may be more important than the level of antioxidant defenses (Orr and Sohal 1992, 1993). More recently, co-overexpression of both enzymes in an isogenic background was found to raise both median and maximum life spans up to 34% (Orr and Sohal 1994; Fig. 7A). This increase was accompanied by decreased mitochondrial H_2O_2 release, decreased protein carbonyl and DNA 8-OHdG accretion, increased resistance to experimental oxidative stress, attenuation of age-related losses of enzyme activity, and decreased attrition of physiological function, as measured by walking speed in negative geotaxis experiments (Sohal et al. 1995a). Finally, as stated above, the metabolic potential of the flies was also increased (Fig. 7B). These results were interpreted as indicating that since Cu-Zn SOD and catalase act in tandem, their overexpression is only effective in tandem. The relative amounts of these enzymes may thus be more crucial than the absolute level of either individual enzyme.

Additional Tests of the Oxidative Stress Hypothesis

The cumulative results of many experiments show increasing levels of oxidative stress and damage during aging in different species. Comparisons within and among species show an inverse relationship between oxidative stress and MLSP. In most cases, the relationships between prooxidant production and aging are more striking than those between antioxidant defenses and aging. The notable exception is the direct effects of Cu-Zn SOD and catalase co-overexpression on life spans and oxidative damage in *Drosophila*. It is possible that co-overexpression of these enzyme neutralizes a greater fraction of ROS before they gain access to certain compartments of the cell, thereby effectively lowering the level of ROS production, rather than raising defenses, from the perspective of those compartments. This interpretation is favored by the observation that mitochondria from transgenic flies with increased life spans release H_2O_2 at a significantly lower rate. A possible explanation may be that one consequence of oxidative damage is structural changes, e.g., in inner mitochondrial membranes, which enhance the rate of ROS production in a positive feedback loop. By lowering the rate of damage from the outset, the overexpressed enzymes may thereby weaken this loop and lower the dose of ROS received by the rest of the cell.

In light of the importance of mitochondrial prooxidant generation, current efforts are directed at bolstering intramitochondrial antioxidant levels, thereby lowering the rate of ROS release to all other parts of the cell. Transgenic lines overexpressing MnSOD have been generated and are currently being characterized. Glutathione peroxidase, which acts in tandem with MnSOD in mitochondria in some systems, has proven refractory in *Drosophila*. The flies do not contain detectable levels of endogenous glutathione peroxidase activity, and attempts to express the human enzyme were thwarted by the inability of *Drosophila* to translate the unique selenocysteine codon at the active site. Recent reports of intramitochondrial catalase activity in rat heart (Radi et al. 1991) raise the possibility that catalase could be modified to enter mitochondria and substitute for glutathione peroxidase in these experiments.

In summary, experiments conducted in this laboratory suggest that oxidative stress may be a fundamental causal factor in aging in representative species of at least two phyla. ROS production increases during aging, and the resulting increase in oxidative stress is associated with an exponential increase in molecular oxidative damage. The rates of prooxidant production and accrual of damage vary in a species-specific manner, indicating that the genetic identity of the animal establishes the level of ROS production, as well as the defensive capability, under a given set of environmental conditions. Thus, contrary to the division of causal theories of aging into separate genetic and environmental classes (Medvedev 1990), the oxidative stress hypothesis emphasizes the interaction between genotype and environment in determining phenotype. The importance of environment is demonstrated by the regimens, such as DR, LA, and temperature reduction, which increase life spans without altering the genotype or the metabolic potential. The importance of genotype is reflected by changes in both life span and metabolic potential in response to overexpression of antioxidant genes under constant environmental conditions. The fundamental importance of oxidative stress is indicated by the observation that all regimens known to increase MLSP also decrease the level of oxidative stress.

Acknowledgments

This research is being supported by grants from the National Institute on Aging – National Institutes of Health (United States Public Health Service).

References

Agarwal S, Sohal RS (1993) Relationship between aging and susceptibility to protein oxidative damage. Biochem Biophys Res Commun 194:1203–1206

Agarwal S, Sohal RS (1994a) Aging and proteolysis of oxidized proteins. Arch. Biochem. Biophys. 309:24–28

Agarwal S, Sohal RS (1994b) DNA oxidative damage and life expectancy in houseflies. Proc. Natl. Acad. Sci U.S.A. 91:12332–12335

Ames BN, Shigenaga MK, Hagen TM (1993) Oxidants, antioxidants, and the degenerative diseases of aging. Proc. Natl. Acad. Sci. U.S.A. 90:7915–7922

Chance B, Sies H, Boveris A (1979) Hydroperoxide metabolism in mammalian organs. Physiol. Rev. 59:527–605

Dubey A, Forster MJ, Lal H, Sohal RS (1996) Effect of age and caloric intake on protein oxidation in different brain regions and on behavioral functions of the mouse. Arch. Biochem. Biophys. 333:189–197

Finch CE (1990) Longevity, senescence and the genome. University of Chicago Press, Chicago

Fries JF (1980) Aging, natural death, and the compression of morbidity. N. Engl. J. Med. 303:130–135

Fucci L, Oliver CN, Coon MJ, Stadtman ER (1983) Inactivation of key metabolic enzymes by mixed-function oxidation reactions: Possible implication in protein turnover and aging. Proc. Natl. Acad. Sci U.S.A. 80:1521–1525

Holmes GE, Bernstein C, Bernstein H (1992) Oxidative and other DNA damages as the basis of aging: a review. Mutat. Res. 275:305–315

Ku H-H, Sohal RS (1993) Comparison of mitochondrial prooxidant generation and antioxidant defenses between rat and pigeon: Possible basis of variation in longevity and metabolic potential. Mech. Ageing Dev. 72:67–76

Ku H-H, Brunk UT, Sohal RS (1993) Relationship between mitochondrial superoxide and hydrogen peroxide production and longevity of mammalian species. Free Radic. Biol. Med. 15:621–627

Medvedev Z (1990) An attempt at a rational classification of theories of aging. Biol. Rev. 65:375–398

Noy N, Schwartz H, Gafni A (1985) Age-related changes in the redox status of rat muscle cells and their role in enzyme aging. Mech. Ageing Dev. 29:63–69

Orr WC, Sohal RS (1992) The effects of catalase gene overexpression on life span and resistance to oxidative stress in transgenic *Drosophila melanogaster*. Arch. Biochem. Biophys. 297:35–41

Orr WC, Sohal RS (1993) The effects of Cu-Zn superoxide dismutase gene overexpression on life span and resistance to oxidative stress in transgenic *Drosophila melanogaster*. Arch. Biochem. Biophys. 301:34–40

Orr WC, Sohal RS (1994) Extension of life-span by overexpression of superoxide dismutase and cata-lase in *Drosophila melanogaster*. Science 263:1128–1130

Pearl R (1928) The rate of living. Alfred A. Knopf, Inc., New York

Radi R, Turrens JF, Chang LY, Bush KM, Crapo JD, Freeman BA (1991) Detection of catalase in rat heart mitochondria. J. Biol. Chem. 266:22028–22034

Richardson A (1991) Changes in the expression of genes involved in protecting cells against stress and free radicals. Ageing Clin. Exp. Res. 3:403–405

Sacher GA (1977) Relation of lifespan to brain weight and body weight in mammals. In: Wolsten-home G, O'Connor M (eds) Ciba Federation Colloquia on Aging, Churchill, London, pp 115–133

Smith CD, Carney JM, Starke RP, Oliver CN, Stadtman ER, Floyd RA, Markesbery WR (1991) Excess brain protein oxidation and enzyme disfunction in normal aging and in Alzheimer's disease. Proc. Natl. Acad. Sci. U.S.A. 88:10540–10543

Sohal RS, Sohal BH (1991) Hydrogen peroxide release by mitochondria increases during aging. Mech. Ageing Dev. 57:187–202

Sohal RS, Orr WC (1992) Relationship between antioxidants, prooxidants, and the aging process. Ann. N. Y. Acad. Sci. 663:74–84

Sohal RS, Dubey A (1994) Mitochondrial oxidative damage, hydrogen peroxide release and aging. Free Radic. Biol. Med. 16:621–626

Sohal RS, Orr WC (1995) Is oxidative stress a causal factor in aging? In: Esser K, Martin GM (eds) Molecular aspects of aging. John Wiley, Chichester, pp. 109–127

Sohal RS, Muller A, Koletzko B, Sies H (1985) Effect of age and ambient temperature on *n*-pentane pro-duction in adult housefly, *Musca domestica*. Mech. Ageing Dev. 29:317–326

Sohal RS, Marzabadi MR, Galaris D, Brunk UT (1989a) Effect of ambient oxygen concentration on lipofuscin accumulation in cultured rat heart myocytes – A novel *in vitro* model of lipofuscino-genesis. Free Radic. Biol. Med. 6:23–30

Sohal RS, Svensson I, Sohal BH, Brunk UT (1989b) Superoxide anion radical production in different animal species. Mech. Ageing Dev. 49:129–135

Sohal RS, Arnold L, Orr WC (1990a) Effect of age on superoxide dismutase, catalase, glutathione reductase, inorganic peroxides, TBA-reactive material, GSH/GSSG, NADPH/NADP⁺ and NADH/NAD⁺ in *Drosophila melanogaster*. Mech. Ageing Dev. 56:223–235

Sohal RS, Arnold LA, Sohal BH (1990b) Age-related changes in antioxidant enzymes and prooxidant generation in tissues of the rat with special reference to parameters in two insect species. Free Radic. Biol. Med. 10:495–500

Sohal RS, Sohal BH, Brunk UT (1990c) Relationship between antioxidant defenses and longevity in dif-ferent mammalian species. Mech. Ageing Dev. 53:217–227

Sohal RS, Svensson I, Brunk UT (1990d) Hydrogen peroxide production by liver mitochondria in dif-ferent species. Mech. Ageing Dev. 53:209–215

Sohal RS, Agarwal S, Dubey A, Orr WC (1993a) Protein oxidative damage is associated with life expec-tancy of houseflies. Proc. Natl. Acad. Sci. U.S.A. 90:7255–7259

Sohal RS, Ku H-H, Agarwal S (1993b) Biochemical correlates of longevity in two closely-related rodent species. Biochem. Biophys. Res. Commun. 196:7–11

Sohal RS, Ku H-H, Agarwal S, Forster MJ, Lal H (1994) Oxidative damage, mitochondrial oxidant gen-eration and antioxidant defenses during aging and in response to food restriction in the mouse. Mech. Ageing Dev. 74:121–133

Sohal RS, Agarwal A, Agarwal S, Orr WC (1995a) Simultaneous overexpression of Cu,Zn superoxide dismutase and catalase retards age related oxidative damage and increases metabolic potential in *Drosophila melanogaster*. J. Biol. Chem. 270:20224–20229

Sohal RS, Agarwal S, Sohal BH (1995b) Oxidative stress and aging in Mongolian gerbil (*Meriones unguiculatus*). Mech. Ageing Dev. 81:15–25

Sohal RS, Sohal BH, Orr WC (1995c) Mitochondrial superoxide and hydrogen peroxide generation, protein oxidative damage, and longevity in different species of flies. Free Radic. Biol. Med. 19:499–504

Stadtman ER (1992) Protein oxidation and aging. Science 257:1220–1224

Stadtman ER, Starke-Reed PE, Oliver CN, Carney JM, Floyd RA (1992) Protein modifications in aging. In: Emerit I, Chance B (eds) Free radicals and aging. Birkhauser, Basel, pp 64–72

Starke-Reed PE, Oliver CN (1989) Protein oxidation and proteolysis during aging and oxidative stress. Arch. Biochem. Biophys. 275:559–567

Weindruch R, Walford RL (1988) The retardation of aging and disease by dietary restriction. Thomas, Charles C, Springfield IL.

Identifying and Cloning Longevity-Determining Genes in the Nematode

T. E. Johnson, S. Murakami, D. R. Shook, S. A. Duhon, and P. M. Tedesco[*]

Abstract

Gerontogenes are genes that are involved in aging. Some alleles of such genes can lead to longer life expectancy. Such genes are beginning to be identified in several species, especially yeast, the fruit fly and in the nematode, *Caenorhabditis elegans*. Five genes are known in the nematode and all of these lead to increased resistance to a variety of stresses. We hypothesize that such genes have been conserved over evolutionary time and serve a role in modulating an organisms response to stress and novel environments.

Introduction

The first papers (Johnson and Wood 1982; Klass 1983; Luckinbill et al. 1984; Rose 1984) showing that it is possible to obtain long-lived metazoan organisms using genetic approaches appeared only some 14 years ago. As a result it can no longer be argued that life span is not under genetic control. This genetic approach has revolutionized the study of aging (Johnson and Lithgow 1992; Curtsinger et al. 1995; Martin et al. 1996). The initial studies cited above used analysis of recombinant inbred strains (RIs), selection of induced mutants and selective breeding strategies, respectively. In the nematode, Caenorhabditis elegans, these initial studies have been followed up by more sophisticated analytic strategies of the RIs including quantitative trait locus (QTL) mapping (Brooks and Johnson 1991; Ebert et al. 1993; Shook et al. 1996) and the cloning of genes, identified as long-lived mutants (see Martin et al. 1996, for an overview), that are involved in life prolongation, the so-called "gerontogenes" (Rattan 1985). We prefer this term rather than the alternative, "longevity assurance genes", which simultaneously implies a mechanism and fails to include many of the genes identified in the nematode whose normal function seems to limit longevity rather than ensure it (Johnson et al. 1996; Martin et al. 1996).

[*] Institute for Behavioral Genetics, University of Colorado, Boulder, Colorado, 80309-0447

Address correspondence and reprint requests to Tom Johnson, Institute for Behavioral Genetics, Box 447, University of Colorado, Boulder, CO 80309

(Phone: 303-492-0279, -7362; FAX: (303) 492-8063; E mail: johnsont@colorado.edu)

J.-M. Robine et al. (Eds.)
Longevity: To the Limits and Beyond
© Springer-Verlag Berlin Heidelberg New York 1997

Three species have been the primary focus of the efforts to identify geronto-genes: Saccharomyces cerevisiae (reviewed by Jazwinski 1996), C. elegans (reviewed by Lithgow 1996), and the fruit fly Drosophila melanogaster (reviewed by Fleming and Rose 1996). The focus on these species arises from two consid-erations: the desire to complete the research in a reasonable period of time, which leads to the use of short-lived species, and the need for sophisticated ge-netic manipulations, which limit the species to be analyzed to these few. Here we will review the literature describing the identification and cloning of these genes in the nematode, C. elegans. Although much of the work is form our laboratory, there has been a recent surge in interest in this area by other investigators work-ing with C. elegans and their work will also be included in this review. For detailed knowledge and background on the nematode system, we suggest Wood (1988).

Several hundred laboratories across the world use C. elegans to study devel-opment, behavior, and physiology. C. elegans reproduces by self-fertilization, yet has a facultative male, which makes it extremely easy to identify mutations, even those that affect life span (Klass 1983; Duhon et al. 1996). For analysis of life extension and other life history traits, the lack of inbreeding depression is extremely important (Johnson and Wood 1982; Johnson and Hutchinson 1993). Moreover, because of its small genome size (10^8 base pairs) and genetic sophis-tication, C. elegans has been chosen as a model organism for the Human Genome Project. The genome has been almost entirely cloned and is available in a variety of contiguous arrays (Coulson et al. 1988); the genome is currently about 60 % sequenced (Wilson et al. 1994) and the entire sequence should be finished by the end of 1998.

Quantitative Approaches to Identifying Gerontogenes

Before a genetic analysis can begin, genetic variants in the phenotype of interest must be identified or constructed. So far two approaches have been used in C. elegans and we will initially describe the first approach, which uses two "wild-type" strains to analyze the genes affecting longevity. This approach takes advan-tage of the inbreeding nature of C. elegans to create many recombinant inbred strains (RIs), each of which is a homozygous strain containing an average 50:50 mixture of genes from each parent. Thus, if the parents differ in any alleles that affect longevity, these allelic variants will be found in differing combinations among the RI strains and these RIs will have life spans that are different from those of the parents. Johnson and Wood (1982) demonstrated a significant ge-netic component affecting longevity. In these initial studies we showed that life span is about 40 % heritable (40 % of the variation in the population was deter-mined by the genes and the remainder by the environment). Some strains had life expectancies almost 70 % longer than those of either parent strain and maxi-mum life spans that were more than two-fold longer than the parents' (Johnson 1987a).

The Genetic Approach

Genetic approaches offer very fine probes for manipulating and identifying processes underlying all aspects of an organism's biology. The major advantages of these approaches are that virtually any gene can be manipulated at will in the species mentioned above; thus, any biochemical process and its involvement in organismic physiology can be studied using an almost endless variety of molecular probes that alter the normal function in the organism. The genetic approach is especially valuable in the analysis of aging because genetic manipulations leading to a stable alteration of genotype also lead to a stable alteration of phenotype in the organism. This phenotypic alteration can now be studied at many levels from the in vivo effects on life expectancy, to altered biochemical and physiological processes and even to studies at the level of the molecule. Also the genetic approach is unbiased by the interests and prior expectations or hypotheses of the investigator in that virtually any gene in the organism that leads to the desired phenotypic change can be identified and studies as a causal determinant of longevity extension. Genetic approaches lead to more than a finding; they lead to an altered strain or stock that can be shared with other laboratories. This strain development is such an important part of the genetic approach that many major journals require the sharing of genetic resources reported within their pages. Finally, perhaps the major advantage of genetic approaches is that they ultimately cross species boundaries, in that evolutionary conservation of gene sequence identify can lead to the identification of homologous genes (genes that retain similar sequence and frequently similar function) in another species. Even the genetic approach has its problems, including unexpected and untoward effects of genetic manipulations such as marker effects that are especially problematic with regard to the study of longevity because life span is affected by so many exogenous factors, both genetic and environmental.

Background on C. elegans

C. elegans is a small, soil-dwelling, non-parasitic nematode worm that has been the subject of many genetic studies during the last 20 years. Brenner (1974) identified numerous morphological and behavioral mutants of C. elegans and established the first genetic maps.

When we looked for evidence of involvement of these genes in specifying other aspects of life history, we found that both development and the length of the reproductive period were independently specified, showing no genetic correlations (Johnson 1987 a). However, the long-lived strains showed prolonged ability to maintain movement, suggesting that longevity and "increased vigor" (at least for a worm) were specified by the same genes. This finding effectively rules out "clock" models that suggest a single timer that controls all temporal aspects of life history (Johnson 1987 b). These initial RIs were supplemented by additional RIs that were used to analyze the relationship between reproduction and longevity, looking for trade-offs between early fertility and longevity as predicted by evolutionary theory; no evidence was found (Brooks and Johnson 1991).

Since then a more detailed analysis of these 81 RI strains has been completed (Shook et al. 1996). We identified quantitative trait loci or "QTLs" specifying longevity and other life history traits in these RI strains. This approach allows us to identify regions of the genome associated with longevity and other life history traits and also to examine epistatic interactions between loci and pleiotropic effects on different traits at specific loci. QTLs for life expectancy were identified on linkage groups (LGs) II (near the stP101 molecular marker), IV (near stP5) and X (near stP61), and QTLs for fertility were identified on LGs II (near maP1), III (near stP19) and IV (near stP51). The QTLs for life expectancy accounted for almost all of the genetic variance (only 23 % of the phenotypic variance). The QTLs for mean fertility accounted for 85 % of the genetic variance and 45 % of the phenotypic variance. Additional QTLs for other life history traits were also mapped in these crosses. There was no evidence for epistatic effects between QTLs but several loci were observed to have effects on the QTLs identified. No QTLs were pleiotropic (having effects on different life history traits) with the exception of negatively correlated, pleiotropic effects between life expectancy and internal hatching associated with a QTL near the stP5 marker.

Mutational Approaches to Identifying Gerontogenes

The best way to identify genes involved in any process is through the use of induced mutants. The search for mutants with increased life span has been tremendously successful in nematodes. Klass (1983) and Duhon et al. (1996) have identified mutants with extended life spans after direct searches of mutagenized F_2 clones. Duhon and Johnson devised a method that eliminated the need to make replicate plates by using a double Daf Ste mutant as the parental strain; at the non-permissive temperature of 25 °C, an asynchronous population of mixed ages of this double mutant produces both sterile adults (Ste) and dauers (Daf). Both are non-reproducing, and the sterile adults can be scored for life span and the mutations recovered from among the Daf siblings.

age-1 was the first mutant in a gerontogene. It was identified by Klass (1983) and subsequently mapped by Friedman and Johnson (1988a, b) and Johnson et al. (1993) to the middle of linkage group II. This mutation behaves as a single-gene and results in a 65 % extension of mean life span and a 110 % extension of maximum life span. The mutant has essentially normal rates of development and fertility. The extended life span is due to a three-fold slowing in the exponential rate of mortality increase (Johnson 1990). Four other mutations lead to significant extension of adult life span in C. elegans. spe-26 mutants result in life extensions of about 65 % for the hermaphrodite and the mated male (Van Voorhies 1992; Murakami and Johnson 1996), although recent observations (Gems and Riddle 1996) suggest that the life extension may be artifactual and result from inappropriate comparisons with wild type. daf-2 mutants result in a more than two-fold extension of mean life span (Kenyon et al. 1993), and this extension is blocked by the action of daf-16. Larsen et al. (1995) showed that daf-23 doubles the life expectancy, and also showed that daf-2 interacts with daf-12 to cause an almost four-fold enhancement of life prolongation. Wong et al. (1995) reported that several alleles of a new gene, clk-1, which have altered the normal course of the cell cycle and of development, also have a increased life span. The several unpublished cases of additional life-extension loci suggest that the total number of gerontogenes in C. elegans may be near 10.

Physiological Role of These Gerontogenes

All of the mutations mentioned above, with the exception of age-1, also alter some other aspect of the normal development or physiology of the nematode (Table 1). spe-26 mutants are male sterile, indicating a defective step in fertilization that turns out to be at the point of spermatid formation (Varkey et al. 1995). The daf-2 and daf-23 mutants are dauer constitutive; they make dauers (an alternate third larval stage that serves as an environment-resistant stage used for dispersal and long-term survival), whereas daf-12 and daf-16 are dauer defective and cannot make normal dauers. At least some of the Daf mutations define a signal-transduction pathway in which homologs to mammalian genes involved in signal transduction can be identified. Recently, T. Inoue and J. Thomas (personal communication) found that age-1 mutants have a second unusual phenotype: dauer formation at 27 °C and an inability to complement daf-23, suggesting that age-1 may be allelic to daf-23. This simple interpretation is complicated by the fact that no age-1 mutant alleles have been found to show mutations in the daf-23 coding sequence (Morris et al., personal communication) and the fact that many other strains also are dauer constitutive at 27 °C (Murakami et al., unpublished). The true identity of age-1 remains unclear.

Both Larsen (1993) and Vanfleteren (1993) showed that age-1 had higher levels of the active oxygen defense systems, SOD and catalase, and was resistant to both paraquat and H_2O_2, relative to wild type. Kenyon et al. (1993) suggested that the role of the gerontogene, daf-2, was to turn on some aspects of the dauer stage, i.e., its life-prolongation phenotype, without turning on the dauer morphology aspects of these dauer genes. Subsequently, the discovery of a similar life-prolongation effect of another dauer gene, daf-23 (Larsen et al. 1995), and the possible involvement of age-1 in the dauer pathway (see above) led us to test this hypothesis. Prior work had shown that age-1, daf-2, and spe-26 were resistant to a variety of thermal stresses (Lithgow et al. 1995). Subsequently, we (Murakami and Johnson 1996; Duhon et al., 1996; Murakami et al., unpublished) have analyzed additional gerontogene mutant strains and have discovered that all of these strains (including age-1, daf-2, daf-23, spe-26, and clk-1) are resistant to a variety of stresses, including H_2O_2, heat, and UV.

UV irradiation has been used in a complete analysis of these gerontogenes and the possible involvement of the dauer pathway in specifying their resistance.

Table 1. Life Span Extension Mutants in C. elegans

Gene name	Life expectancy[1] % increase	(Ref)[2]	Other Phenotypes[1]	Resistant to[1] ROS	Heat	UV	(Ref)	Suppression by daf-16 Life	ROS	UV	(Ref)	Hsp 16 Elevation	(Ref)
age-1	65	1	27 °C: dauer constitutive Increase SOD and catalase	+	+	+	10, 11, 12 10	+	+	+	5, 7, 12	++	14
spe-26[3]	65	2	Ts sterile	+	+	+	7, 11, 12	+	+	+	7, 12	++	14
daf-2	100	3	25 °C: dauer constitutive	+	+	+	7, 11, 12	+		+	3, 12	++	14
clk-1[3]	40	4	Cell cycle altered	+	+/-	+	7, 12, 15	+	+	+	7, 12	+/-	15
daf-23	100	5	25 °C: dauer constitutive		+	+	7, 12, 16	+		+	7, 12		
daf-12	300 (with daf-2)	5											
daf-28	30 dominant	6	Ts sterile		+		13						
spe-10	40	7									7		
age-n(z11)	40	8	Reduced fertility?				7						
age-n[4]	40	8, 9					7, 9						

[1] +, resistant; O, no effect; blanks have not yet been tested.

[2] References: 1. Klass 1983; Friedman and Johnson 1988a; Johnson 1990. 2. Van Voorhies 1992. 3. Kenyon et al. 1993. 4. Wong et al. 1995. 5. Larsen et al. 1995. 6. T.I-noue and J. Thomas, personal communication. 7. S. Duhon and T. E. Johnson, unpublished. 8. Duhon et al. 1996. 9. Y. Yang and D. Wilson, personal communication. 10. Vanfleteren 1993; Larsen 1993. 11. Lithgow et al. 1995. 12. Murakami and Johnson 1996. 13. G. Lithgow, unpublished. 14. Lithgow et al., submitted. 15. P. Tedesco and T. E. Johnson, unpublished. 16. S. Dale and T. E. Johnson, unpublished.

[3] These genes showed allelic variation with only some alleles showing life extension; all life-extending alleles tested were also UV resistant.

[4] Newly isolated mutants fail to complement age-1; other tests still pending.

Mutations in daf-16 suppressed the UV resistance as well as the increased longevity of all of these mutants (Murakami and Johnson 1996). In contrast to the many physiological mechanisms proposed initially, these findings suggest that a single pathway, involving daf-16, specifies both the life-extension and UV-resistance phenotypes. The fact that mutations in daf-16 did not alter the reduced fertility of spe-26 (interestingly, a daf-16 mutant is more fertile than wild type) but still reduced life span suggests conclusively that fertility is not the determinative event for life span determination. A model in which both life span and some stress-resistance pathways are jointly negatively regulated by these gerontogenes is consistent with these results.

We (Lithgow et al. 1995) tested the causal connection between increased thermotolerance and increased life span by developing conditions for environmental induction of thermotolerance. Treatment at sub-lethal temperatures induced large increases in thermotolerance; small, but statistically highly significant, increases in life expectancy were also observed, consistent with a causal connection between heat tolerance and life span. A larger variety of additional stress-resistance mechanisms could have also been induced by these treatments and this currently is under investigation (Murakami and Johnson, unpublished). These results show that when an organism's resistance to stress is increased, by either genetic or environmental manipulation, an increase in life expectancy can result. These findings suggest that the ability to respond to stress limits life expectancy in C. elegans and suggest that similar processes may also limit life span in other metazoa.

Cloning the Gerontogenes

Only two of the gerontogenes mentioned above have been cloned: spe-26 and daf-23. These two genes code for a protein similar to the actin-associated proteins, kelsh and scruin (Varkey et al. 1995), and a phosphoinositol-3 kinase homolog (Morris et al., personal communication), respectively. Both gerontogenes were cloned using the integrated physical genetic map and data base available for C. elegans to identify cosmids in the correct genetic region followed by complementation analysis to show that the cloned gene could correct the deficit. These are standard cloning practices for C. elegans. Following the complementation tests, sequence analysis is used to confirm the presence of altered sequence in each of the mutant strains.

References

Brenner S (1974) The genetics of Caenorhabditis elegans. Genetics 77:71–94
Brooks A, Johnson TE (1991) Genetic specification of life span and self-fertility in recombinant-inbred strains of Caenorhabditis elegans. Heredity 67:19–28
Coulson A, Waterston R, Kiff J, Sulston J, Kohara Y (1988) Genome linking with yeast artifical chromosomes. Nature 335:184–186

Curtsinger JW, Fukui HH, Xui L, Khazaeli AA, Kirscher A, Pletcher SD, Promislow DEL, Tatar M (1995) Genetic variation and aging. Ann Rev Genet 29:553–575

Duhon SA, Murakami S, Johnson TE (1996) Direct isolation of longevity mutants in the nematode Caenorhabditis elegans. Develop Genet 18:144–153

Ebert RH, Cherkasova VA, Dennis RA, Wu JH, Ruggles S, Perrin TE, Shmookler-Reis RJ (1993) Longevity-determining genes in Caenorhabditis elegans: Chromosomal mapping of multiple noninteractive loci. Genetics 135:1003–1010

Fleming JE, Rose MR (1996) Genetics of aging in Drosophila. In: Rowe JW, Schneider EL (eds) Handbook of the biology of aging. 4th edn. Academic Press, New York, pp 74–93

Friedman DB, Johnson TE (1988a) A mutation in the age-1 gene in Caenorhabditis elegans lengthens life and reduces hermaphrodite fertility. Genetics 118:75–86

Friedman DB, Johnson TE (1988b) Three mutants that extend both mean and maximum life span of the worm, Caenorhabditis elegans, define the age-1 gene. J Gerontol Bio Sci 43:B102–B109

Gems D, Riddle DR (1996) Mating but not gamete production reduces longevity in Caenorhabditis elegans. Nature 379:723–725

Jazwinski SM (1996) Longevity-assurance genes and mitochondrial DNA alterations: Yeast and filamentous fungi. In: Rowe JW, Schneider EL (eds) Handbook of the biology of aging. 4th edn. Academic Press, New York, pp 39–54

Johnson TE (1987a) Aging can be genetically dissected into component processes using long-lived lines of Caenorhabditis elegans. Proc Natl Acad Sci USA 84:3777–3781

Johnson TE (1987b) Developmentally programmed aging: Future directions. In: Warner HR (ed) Modern biological theories of aging. Raven Press, New York, pp. 63–76

Johnson TE (1990) The increased life span of age-1 mutants in Caenorhabditis elegans results from lowering the Gompertz rate of aging. Science 249:908–912

Johnson TE, Hutchinson EW (1993) Heterosis for life span and other life history traits in Caenorhabditis elegans. Genetics 134:463–474

Johnson TE, Lithgow GJ (1992) The search for the genetic basis of aging: the identification of gerontogenes in the nematode Caenorhabditis elegans. J Am Ger Soc 40:936–945

Johnson TE, Wood WB (1982) Genetic analysis of the life-span of Caenorhabditis elegans. Proc Natl Acad Sci USA 79:6603–6607

Johnson TE, Tedesco PM, Lithgow GJ (1993) Comparing mutants, selective breeding, and transgenics in the dissection of aging processes of Caenorhabditis elegans. Genetica 91:65–77

Johnson TE, Lithgow GJ, Murakami S, Duhon SA, Shook DR (1996) Genetics of aging and longevity in lower organisms. In: Holbrook N, Martin GR, Lockshin RA (eds) Cellular aging and cell death. John Wiley and Sons, New York, pp 1–17

Kenyon C, Chang J, Gensch E, Rudner A, Tabtiang R (1993) A C. elegans mutant that lives twice as long as wild type. Nature 366:461–464

Klass MR (1993) A method for the isolation of longevity mutants in the nematode Caenorhabditis elegans and initial results. Mech Ageing Dev 22:279–286

Larsen PL (1993) Aging and resistance to oxidative stress in Caenorhabditis elegans. Proc Natl Acad Sci USA 90:8905–8909

Larsen PL, Albert PS, Riddle DLL (1995) Genes that regulate both development and longevity in Caenorhabditis elegans. Genetics 139:1567–1583

Lithgow GJ (1996) The molecular genetics of Caenorhabditis elegans aging. In: Rowe JW, Schneider EL (eds) Handbook of the biology of aging. 4th edn. Academic Press, New York, pp 55–73

Lithgow GJ, White TM, Melov S, Johnson TE (1995) Thermotolerance and extended life-span conferred by single-gene mutations and induced by thermal stress. Proc Natl Acad Sci USA 92:7540–7544

Luckinbill LS, Arking R, Clare MJ, Cirocco WC, Muck SA (1984) Selection for delayed senescence in Drosophila melanogaster. Evolution 38:996–1003

Martin GM, Austad SN, Johnson TE (1996) Genetic analysis of aging: Role of oxidative damage and environmental stresses. Nat Genetics 13:25–34

Murakami S, Johnson TE (1996) A genetic pathway conferring life extension and resistance to UV stress in Caenorhabditis elegans. Genetics 143:1207–1218

Rattan SIS (1985) Beyond the present crisis in gerontology. Bioessays 2:226–228

Rose MR (1984) Laboratory evolution of postponed senescence in Drosophila melanogaster. Evolution 38:1004–1010

Shook D, Brooks A, Johnson TE (1996) Mapping quantitative trait loci specifying hermaphrodite survival or self fertility in the nematode Caenorhabditis elegans. Genetics 142:801–817

Vanfleteren JR (1993) Oxidative stress and aging in Caenorhabditis elegans. Biochem J 292:605–608

VanVoorhies WA (1992) Production of sperm reduces nematode lifespan. Nature 360:456–458

Varkey JP, Muhlrad PJ, Minniti AN, Do B, Ward S (1995) The Caenorhabditis elegans spe26 gene is necessary to form spermatids and encodes a protein similar to the actin-associated proteins kelsh and scruin. Genes Devel 9:1074–1086

Wilson R, Ainscough R, Anderson K, Baynes C, Berks M, Bonfield J, Burton J, Connell M, Copsey T, Cooper J, Coulson A, Craxton M, Dear S, Du Z, Durbin R, Favello A, Fraser A, Fulton L, Gardner A, Green P, Hawkins T, Hillier L, Jier M, Johnston L, Jones M, Kershaw J, Kirsten J, Laisster N, Latreille P, Lightning J, Lloyd C, Mortimore B, O'Callaghan M, Parsons J, Percy C, Rifken L, Roopra A, Saunders D, Shownkeen R, Sims M, Smaldon N, Smith A, Smith M, Sonnhammer E, Staden R, Sulston J, Thierry-Mieg J, Thomas K, Vaudin M, Vaughan K, Waterston R, Watson A, Weinstock L, Wilkinson-Sproat J, Wohldman P (1994) 2.2 Mb of contiguous nucleotide sequence from chromosome III of C. elegans. Nature 368:32–38

Wong A, Boutis P, Hekimi S (1995) Mutations in the clk-1 gene of Caenorhabditis elegans affect developmental and behavioral timing. Genetics 139:1247–1259

Wood WB (1988) The biology of Caenorhabditis elegans. Cold Spring Harbor Press, New York

Longevity: Is Everything Under Genetic Control?
An Inquiry Into Non-genetic
and Non-environmental Sources of Variation

*C. E. Finch**

Summary

In this essay, I inquire about little explored sources of non-genetic factors in individual life spans that are displayed between individuals with identical genotypes in controlled laboratory environments. The numbers of oocytes found in the ovaries of inbred mice, for example, show a >5-fold range between individuals. Smaller, but still extensive variations are also indicated for hippocampal neurons. These variations in cell number can be attributed to stochastic processes during organogenesis, i.e. "developmental noise in cell fate determination." They may be of general importance to functional changes during aging, as argued for reproductive senescence in females which is strongly linked to the time of oocyte depletion. More generally, I hypothesize that variations in cell numbers during development result in individual differences in the reserve cell numbers which, in turn, set critical thresholds for dysfunctions and sources of morbidity during aging. Thus, future studies on the nongenetic causes of variations in longevity could address developmental noise leading to variations in cell numbers, the external pre- and postnatal environments, and their interactions with the genome throughout the life span.

Introduction

Human populations show a broad range of individual life spans 30–50 % about the mean (Jones 1956, Gavrilov and Gavrilova 1991) that undoubtedly reflects influences from both genotype and environment. However, review of familiar data on laboratory rodent survival curves raises a surprising question: why do laboratory populations of highly inbred mice display a range of individual life spans (Fig. 1) that is similar to that found in genetically polymorphic human populations? I suggest that individual life spans in populations are subject to a little recognized category of nongenetic factors that are displayed in populations of organisms with identical genotypes even when the environment is as strictly

* Neurogerontology Division, Andrus Gerontology Center and Department of Biological Sciences, University of Southern California, Los Angeles CA 90089–0191, 2137401758 phone, 2137400853 FAX

J.-M. Robine et al. (Eds.)
Longevity: To the Limits and Beyond
© Springer-Verlag Berlin Heidelberg New York 1997

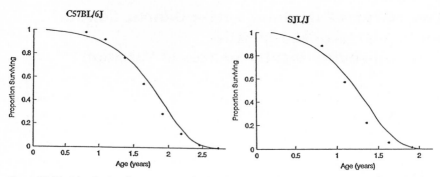

Fig. 1. Highly inbred mice show a range of individual life spans that range 30–50 % about the mean values. Two strains, C57BL/6J, that differ 25 % in mean and maximum life span are shown from a larger set analyzed by Finch and Pike (1996). In the 20 years since the original study was completed, the life spans of each of these strains has increased by about 25 %, which is most likely due to improvements of husbandry rather than genetic drift.

controlled as laboratory conditions allow. This essay explores how these variations may arise epigenetically. I propose the hypothesis that individual variations in cell numbers differentiate individuals by the numbers of reserve cells which set critical thresholds for viability during aging.[1]

Among the biological explanations for these variations in the life spans of individual mammals and other vertebrates is an outcome of cell fate determination during embryonic development, through which the numbers of cells normally show wide variations between individuals. In mammals, like other vertebrates examined, embryogenesis is characterized by the conditional specification of cell fate (Winklbauer and Hausen 1983; Jacobsen 1985; Davidson 1991). That is, the fate of daughter cells in any cell lineage is subject to statistical rules that set the probability for survival as a particular future differentiated cell type, or for cell death during organogenesis that may be considered as "development noise in cell fate determination." This stochastic feature of cell fate assignment arises after the extensive migration of cells that is characteristic of vertebrate and insect embryos, as well as those in certain other phyla (Davidson 1991). Consequently, as described below, there are extensive variations in the numbers of cells between individuals in a given organ. Developmental biologists have long recognized fluctuations in asymmetry that may be another outcome of this development noise, e.g., in the numbers of veins on the wings of flies on each side of the body, or the numbers of teeth on each side of the mouth in mice (Phelan and Austad 1994), but these phenomena have been little discussed in the context of aging and longevity. First I illustrate these phenomena with examples from ovarian oocytes that give a model for thinking about how individual variations in cell number in the nervous system and other organs can be determinants of longevity.

---

[1] These concepts have been further evolved from earlier considerations (Finch 1982, 1990, 1994, 1996; vom Saal et al. 1994).

Cell Numbers Vary Between Individuals due to Developmental Mechanisms

The Ovary

A striking example of variations in cell number is that inbred mouse strains show a > 5-fold variability between individuals in the numbers of oocytes and primary follicles in the ovary at birth and at later ages. Figure 2 shows these variations for C57BL/6J mouse strain, in a study using mice obtained from the Jackson Laboratory. Similar variations between individuals in the numbers of ovarian oocytes have been documented in every report of oocyte number in inbred rodents (Faddy et al. 1983; Jones and Krohn 1961; Gosden et al. 1983; vom Saal et al. 1994). Moreover, a similar range of oocyte numbers is reported in neonatal humans (Block 1952). This wide range of variation in ovarian oocyte pools cannot be readily attributed to genetic variations in the examples of inbred laboratory rodents.[2]

Despite these caveats, there is substantive evidence that the variations in oocyte and follicle number in rodents and other mammals are a result of dispersion and cell death during the migration of primordial germ cells within the embryo over relatively large distances during organogenesis. The primordial germ cells eventually collect to form the gonads. Not all of these cells reach their destination and stragglers can be identified as ectopic germ cells that mostly disappear before puberty (Upadhyay and Zamboni 1982).

An explanation of variable numbers of oocytes can be found in the statistical feature of developmental mechanisms that determine cell lineages and cell type. Cell fate during organogenesis may depend on local variations in trophic molecules. For example, primordial germ cell survival is enhanced by at least three growth factors: SCF, LIF and bFGF (Pesce et al. 1993; Matsui et al. 1992). Besides putative fluctuations in growth factors in the extracellular environment, another cause may be fluctuations in the external microstructure of contiguous cells which mediate cell-cell signaling and production of paracrine-acting growth fac-

---

[2] Nonetheless, it remains possible that some of the oocyte number variation arises through residual polymorphisms. Inbred mouse strains differ significantly in oocyte numbers and on the rate of their loss during aging. For example, in comparisons of the inbred CBA mouse with other strains, the CBA females have a smaller oocyte stock, a faster rate of depletion, and an earlier onset of age-related infertility (Faddy et al. 1983). The genes involved have not been identified, but might be resolved through the growing inventory of recombinant-inbred strains. In regard to humans, we may anticipate that genetic polymorphisms will be identified that influence the age of menopause, through either the initial size of the oocyte pool or the rate of loss during postnatal aging. In the examples of inbred rodents, despite inbreeding to the asymptotic approach of genetic homogeneity in brother-sister mating systems (Klein 1975), it is possible that some genetic heterogeneity remains. Surveillance for genetic homogeneity, of course, emphasizes certain genes that are of operational importance to biologists, e.g., those in the Mhc that determine transplantation antigens. The numbers of endogenous retroviruses may also vary between populations and can interact with the reproductive system, in the case of mouse mammary tumor virus (MMTV).

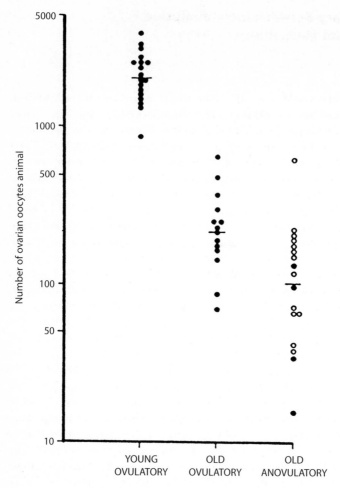

Fig. 2. Inbred C57BL/6J mice show five-fold variability between individuals in the numbers of oocytes at birth. (Redrawn from Gosden et al. 1983) The cycling status was evaluated by daily vaginal cytology. The ovulatory status was confirmed by ovarian histology. Young mice were 4- to 5-months old; old mice were 13 to 14 months; "ovulatory" was defined by continuing estrous cycles and mature ova; "old anovulatory" represents the absence of estrous cycles and of ovulation; open circles (old anovulatory), mice with vaginal cytology indicative of constant estrus (persistent vaginal cornifications, PVC); closed circles, a subpopulation of old mice with irregular vaginal cytology typical of irregular anovulatory cycles that typically precede PVC (see vom Saal et al. 1994).

tors. The multiplicity of growth factors that influence germ cell survival allows for combinatorial influences that are consistent with the observed statistics. Similar mechanisms probably pertain to survival of neurons and many other cell types. These and other features of the micro-environment that determine cell fates are a major frontier in developmental biology.[3]

Whatever the origins of these development variations in oocyte number, there is good evidence that, at later ages, the numbers of surviving cells is a major factor in the age of reproductive senescence in female rodents. In Figure 2, one can see the wide individual variation at age 13–14 months in the numbers of surviving oocytes and in the presence of ovulatory cycles. Below a threshold of approximately 150 surviving oocytes, ovulatory cycles cease and fertility is lost in mice (Gosden et al. 1983; Nelson and Felicio 1986). Women also show individual variations at middle age in the numbers of surviving oocytes and follicles. Wheras chronological age is a weak predictor of the numbers of remaining oocytes, the event of menopause is highly predictive of the complete exhaustion of ovarian oocytes and the hormone-producing follicles (Richardson et al. 1987).

These data from two species imply that there is a correspondence between the numbers of oocytes present at birth and the age of ovarian depletion. Thus, one of the most vital parameters of life history, the age when fertility is lost at menopause in humans, appears subject to considerable non-genetic variability (Faddy and Gosden, 1995). It would be valuable to know the degree of concordance shown by identical twins in the age of menopause.

The argument here can be pressed beyond correlation by an innovative experiment of Nelson and Felicio (1986) that tested the relationship between the numbers of oocytes present in the young ovary to the time of later life reproductive senescence. This study diminished the reserve oocyte pool in young adults by "radical ovarian resection", or surgical removal of most of the ovary. The rate of loss of ovarian oocytes during postnatal aging approximates zero-order kinetics, like radioactive decay, and can be fitted nicely to the linear regression of log (oocytes) vs. age (Jones and Krohn 1961; Gosden et al. 1983; Nelson and Felicio 1986; Fig. 3). From data (Fig. 2) that mice cease cycling during aging when their oocyte and primary follicular reserve is <150, it can be calculated from the regression equation that a 90% reduction of oocyte loss by surgical resection should cause the acceleration of reproductive senescence by about seven months.

---

[3] Here another caveat must be considered, a different possible epigenetic environmental phenomenon from the fetal neighbor effect. Rodents and probably other polytocous species show intra-uterine interactions between adjacent fetuses of different genders that, through influences on fetal sex steroid levels, have important outcomes on reproductive adult performance, including aging (vom Saal et al. 1994, pp 1272-4). For example, female mice that are flanked by males show an earlier onset of reproductive senescence than observed in females that are flanked by females, in association with delayed parturition (vom Saal and Moyer 1985) that increases still-births at later maternal ages in mice (Holinka et al. 1979; vom Saal and Moyer 1985). Although delayed parturition as a cause of reproductive senescence may be neuroendocrine, it remains to be investigated if there are other fetal neighbor effects that influence the surviving numbers of ovarian oocytes though sex steroid levels within the fetal circulation.

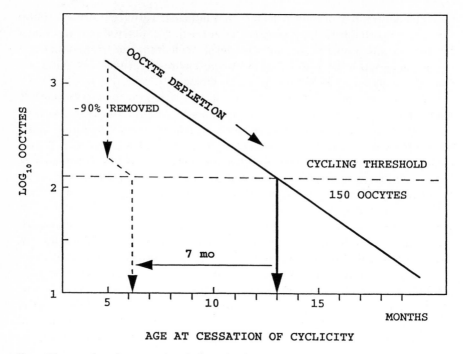

Fig. 3. Diagram of ovarian oocyte loss during aging in C57BL/6J mice, plotted on a semi-logarithmic scale. The surgical removal of 90% of the oocytes at five months accelerated the age when ovulatory cycles were lost by about seven months. (Redrawn from Nelson and Felicio 1986).

The results from longitudinal daily observations of vaginal cytology confirmed this prediction for both the age-related lengthening of cycles as well as the onset of acyclicity (Fig. 3). Thus, it may be hypothesized that variations in the numbers of oocytes that arise before birth in humans are a key factor in the age of natural menopause five decades later. In the case of humans, intrauterine variations of sex steroid levels may be less substantive than in the examples of polytocous rodents, as discussed above.

Menopause in humans is well known for sharp increases of several major risk factors of morbidity and mortality, particularly osteoporosis, with subsequent increased risks of disabling fractures of bone and changes in blood lipids that accelerate atherogenesis with increased risks of heart attacks and strokes. In these conditions, estrogen replacement therapy (ERT) is being well documented to reduce the associated risks. A further recent addition to the list of menopause-associated causes of mortality is Alzheimer's disease in women, in which a lower risk is associated with a history of ERT (Paganini-Hill and Henderson 1994). Even in this currently intractable disease, ERT enhances responses to the cholinergic medication tacrine (Schneider et al. 1996). Because the onset of the loss of ovarian estrogen production at menopause is tightly linked to the exhaustion of the estrogen-producing ovarian follicles, individual variations in the timing of menopause and in the timing of the subsequent estrogen-dependent disorders can thus be traced back to stochastic variations in the numbers of oocytes and follicles present at birth. Thus, variations in the number of oocytes and follicles at birth are likely to be a major influence on diseases that shorten the life span, some of which can be modified by ERT. At this time, however, there is no general correlation between the life span and the timing of reproductive senescence in mammals.

The Down syndrome can be discussion in relation to ovarian oocyte numbers because its risk accelerates with maternal age and, just before menopause, occurs in excess of 10% of all live births (Hook 1981; Gosden 1985; Finch 1990). The Down syndrome represents the majority of maternal-age associated chromosomal aneuploidies. For reasons that are poorly understood, all individuals with chromosome 21 trisomy eventually develop the neuropathology of Alzheimer's disease, with marked cognitive declines in middle age. A link of dementia-inducing aneuploidy to ovarian oocyte numbers is suggested by a powerful experiment in mice by Gosden and colleagues: the incidence of age-related fetal aneuploidy was increased by reducing the oocyte and follicle pool by 50% through removal of one ovary (Eichenlaub-Ritter et al. 1988). In the context of the preceding discussion, individual differences in oocyte numbers at birth would be predicted to influence the risk of fetal aneuploidy at a given age. Because those with the Down syndrome have shortened life expectancy (heart

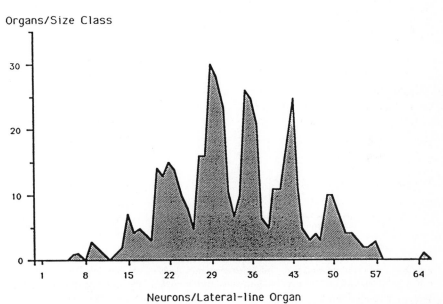

Fig. 4. Neuron number in the frog lateral line system varies between locations in the same individual over a 10-fold range. (Redrawn from Winklbauer and Hausen 1983). The survival of neurons in the lateral line system is determined by a series of binary choices with a fixed probability, like a string of coin tosses (heads, the cell wins and survives; tails, the cell dies). The resulting distribution of neurons in the lateral line fits the binomial distribution.

defects, Alzheimer's disease etc.), stochastic variations in oocyte number could influence the life expectancy of the next generation of this particular genetic sub-type.

The nervous system

Neural tissues also present clear examples of, "developmental noise in cell fate determination" which leads to varying numbers of neurons, not only between individuals, but also within an individual. Certain parts of the nervous system clearly show a fixed probability at each division that one of the daughter cells will survive as a neuron. A striking example is in the lateral line system of frogs, within which the numbers of neurons vary between segmental locations in the same individual over a 10-fold range (Winklbauer and Hausen 1983; Fig. 4). In this system, the survival of neurons is determined by a series of

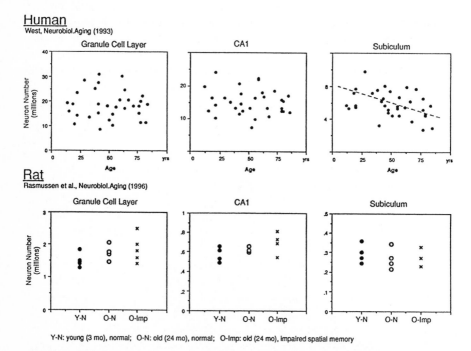

Y-N: young (3 mo), normal;   O-N: old (24 mo), normal;   O-Imp: old (24 mo), impaired spatial memory

Fig. 5. The numbers of neurons vary widely between individuals in two regions of the hippocampal formation of human and rat brain. (The figure is redrawn from West 1993 and Rasmussen et al. 1996). The brains of humans were selected to eliminate individuals with stroke or Alzheimer's disease. The anatomically contiguous subiculum shows a significant trend for neuron loss in human but not rat brain. The 24-month-old rats are functionally divided into two groups: O-N, which were indistinguishable from 3-month-old young rats (Y-N) in a spatial memory test using the Morris water maze; and O-IMP, which showed impairments on this test. Overall, there is a general stability of neuron number during normal aging.

binary choices with an apparently constant probability. Like a string of coin tosses, if heads, the cell wins and survives, if tails, the cell dies or changes its phenotype. The resulting distribution of neurons in the lateral line fits the binomial distribution. This situation contrasts with that in the worm Caenorhabditis and certain other invertebrates, in which the lineage of every somatic cell in the embryo and adult and the numbers of cells in adult Caenorhabditis are invariant. The rigid process of cell determination in Caenorhabditis determines precisely which cells become neurons, for example, and which cells die.

These same principles of stochastic cell determination found in the lateral line system of frogs may also prevail in the vertebrate brain. In mammals and other vertebrates, it is well established that many more neurons are produced during development than survive in the adult, as presented in the classical studies of Victor Hamburger, Ronald Oppenheimer, and others in the spinal cord, in which non-innervated neurons do not survive. The cerebral cortex, furthermore, develops through the dispersal of clonally related cells by migration within a generalized protocortical map that does not precisely specify where each neuron will reside or what its connections will be (e.g. Edelman 1987; Jacobsen 1985). It is possible that variations in neuron number are the basis for 5–40 % of differences between twins in the areas of particular cortical gyri and in the left-right asymmetry (Jouandet et al. 1989).

The question of individual variations in neuron number cannot be easily evaluated from early published papers because of confounds that have been addressed by Mark West, who developed a rigorous technique of optical dissection to minimize double counting of neurons (West et al. 1991; West 1993; Rasmussen et al. 1996). Figure 5 shows data from three regions of human and rat brain, the granule neuron layer and CA1 pyramidal layer of the hippocampus and the contiguous subiculum. The specimens analyzed in Figure 5 are from clinically normal individuals, i.e., without evidence of strokes, Alzheimer's disease, or other gross neurodegeneration. The hippocampal formation is of major importance for learning and is highly vulnerable to neuron loss in Alzheimer's disease and stroke. Two features of neuronal numbers are remarkable.

First, note the range of neuron numbers in each subregion, which in humans appears to span up to 50 % of the mean at all adult ages, whereas in rats the range is less, about 10–20% about the mean in number in young individuals, with coefficients of variation for neuron number of 10–15 % (West et al. 1991). This range of individual numbers of a particular cell type may be less than the >2-fold range found in ovarian oocytes, but is still considerable. Another analysis of healthy humans displays a similar range of variations between individual adults (Henderson et al. 1980).

Second, these data challenge the long-standing belief that neuron loss during aging is extensive. On the contrary, in these disease-free brains, one is struck by the overall stability of neuron numbers across the life span in the absence of stroke or Alzheimer's disease, a possibility that is slowly becoming accepted (e.g., Wickelgren 1996). Thus, for a neuron or other cell type to be merely nondividing does not predestine high age-related cell death rates, unlike the situation

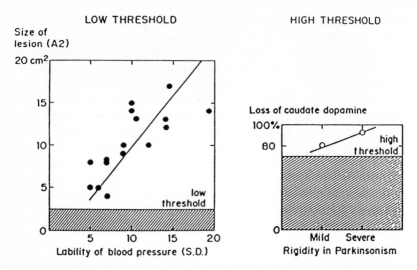

Fig. 6. Relationships between the size of brain lesions and the functional consequences in two types of dopaminergic neurons. Left, the A2 brain stem nucleus shows a low threshold (<20% loss) for functional damage, in which the size of the lesion correlates with the fluctuations (standard deviations, S. D.) of blood pressure. (From Talman et al. 1980). Right, in contrast, the substantia nigra projections show a much higher threshold (>80% loss) before functional manifestations, which in this case is the degree of rigidity in Parkinsonian patients. (From Bernheimer et al. 1973. Redrawn from Finch 1982 and Finch 1990, p. 277)

in the ovary. Although the subiculum shows progressive reduction of neurons at later ages in humans, note that some individuals at later ages clearly have as many neurons as the young. The 2-year-old rats shown in Figure 5 were grouped by learning performance, yet did not show any correlation with neuron number (Rasmussen et al. 1996; see Fig. 5, legend).

The situation in neurodegenerative diseases of aging is sharply different especially in Alzheimer's disease, in which the hippocampus is a major target of neuron loss (Braak and Braak 1991). For neuron populations that are at risk for loss, the numbers of neurons in the hippocampus, for example, might set a threshold for loss that can be incurred without dysfunction. This argument is based on data from other brain regions, in which the extent of neuronal loss has region-specific thresholds for functional consequences. A region with a low threshold for neuron loss is the brain stem, where the size of experimental lesions of the dopaminergic A2 neuron group that regulate blood pressure caused proportionate fluctuation in blood pressure (Talman et al. 1980; Fig. 6). In contrast, the substantia nigra shows a much higher threshold for neuron loss before motor functions are compromised, as observed in Parkinsonism, where movement disorders typically ensue only when >80% of dopaminergic neurons are lost in the substantia nigra pars compacta (Bernheimer et al. 1973).

Puberty may represent another example of these phenomena. In humans, the age of menarche varies widely among individuals for reasons that are poorly understood. Menarche, or the onset of ovarian hormonal cycles at puberty, is thought to be controlled by one or more neural loci that are sensitive to some hormonal or metabolic correlate of growth, that are referred to as a "sommato-meter" (Plant 1994). I have suggested that variations in the age of puberty of identical monozygous twins could arise from variations in the numbers of neurons in the brain loci that determine the onset of puberty; wheras, as discussed above, variations among twins in the age of menopause could arise from variations in oocyte and follicles (Finch 1996). There are no data on the concordance of ages at menarche and menopause in mono- and dizygous twins.

Other Organisms

Among the commonly used animal models for studies of aging are inbred fruit flies and nematodes. The fruit fly Drosophila, like other arthropods, also share with the cordates the developmental trait of statistical fluctuations of cell fate that are likely to lead to wide individual variation in cell number. In contrast, the nematode Caenorhabditis also shows considerable nongenetic variation in life span (see Johnson et al., this volume). Although the numbers of cells are very strictly controlled in Caenorhabditis, nonetheless the connections between neurons that determine variations in function may be somewhat independent of neuron cell numbers. The origin of these nongenetic variations in mortality statistics is obscure, but must have a different nature than that found in mammals.

Conclusions

These biological perspectives suggest that there are multiple sources of nongenetic variance in life spans. The evidence reviewed suggests the need for another term for the standard equation for sources of phenotypic variance that represents stochastic (unprogrammed) variations in cell number, V_{SVCN}. Stochastic variations in cell number are subject to genotype (G) and environmental (E) factors, but should be distinguished from $G \times E$ interactions that more typically represent external environmental effects on gene expression.

$$V_{phenotype} = V_G + V_E + V_{SVCN}$$

This equation is merely hypothesis and could be expanded to include interactions between the terms, e.g., between V_{SVCN} and V_G, as implied for the mouse strain differences in the rat of ovarian oocyte loss during aging (Jones and Krohn 1961). I propose that variance in life spans can also be resolved according to this formulation. In particular, I hypothesize that the variance in life spans observed in inbred mouse and rat genotypes is a result of V_E from intrauterine phenomena (see footnote 3) that lead to different social interactions as adults (vom Saal et al. 1994), but also from V_{SVCN} that yield different factors of safety in the reserve function of the kidney, an organ that is highly vulnerable to deterioration during aging in rodents (Bronson 1990).

Although allelic variations in genes clearly influence the numbers of cells in adult vertebrates, the genetic determinants appear to set the range of cell numbers for any type of cell, rather than an absolute number. These variations could be key substrates of individual variations of life history. These individual variations in the ab initio reserves of irreplaceable oocytes are plausible determinants of the age when fertility is lost due to exhaustion of the oocyte pool. Future analyses may indicate if the age of menopause is a factor in subsequent life expectancy, as would be predicted from the association with two causes of morbidity, osteoporosis and vascular disease, and the long-term impact of estrogen replacement therapy (ERT) on life expectancy. The survival of Jeanne Calment to 122 years without ERT indicates that other factors can ameliorate the loss of estrogens.

The non-genetic sources of variation during development could account for some of the variance in life spans by the size of the reserve cell pools in different organs. In the nervous system, brain regions differ widely in the threshold of neuron loss beyond which major dysfunctions emerge. With the many studies of neuron number that are ongoing, we may anticipate a brain map that gives the quantitative variations in different neuron populations and the thresholds of neuron loss during age-related neurological disease that are associated with different degrees of functional change. Some populations of neurons may prove to be strongly linked to life expectancy.

Similar questions can be posed for the immune system, which depends on clonally lineages of cells that are derived from obscurely enumerated stem cells. A possible consequence is variations between individuals in immune responses at later ages that could be a determinant of resistance to infection or the proclivities to autoimmune disease. Thus it is likely that organ systems differ widely during aging between individuals in the stochastically determined numbers of cells above critical thresholds for irreversible loss of function that influence morbidity and mortality.

Acknowledgements

I am grateful for comments made by critical readers of this manuscript (Julie Andersen, Pamela Larsen, Todd Morgan of USC, and Mark West of the Institute of Anatomy at Aarhus, Denmark). This work was also supported by grants from the NIA (AG 09793; AG 00729)

References

Bernheimer H, Birkmayer W, Hornykiewicz O, Jellinger K, Seitelberger F (1973) Brain dopamine and the syndromes of Parkinson and Huntington: Clinical, morphological, and neurochemical correlations. J Neurol Sci 20:415–455

Block E (1952) Quantitative morphological investigations of the follicular system in women. Acta Anat 14:108–123

Braak H, Braak E (1991) Neuropathological stageing of Alzheimer-related changes. Acta Neuropathol 82:239–259

Bronson RT (1990) Rate of occurrence of lesions in 20 inbred and hybrid genotypes of rats and mice sacrificed at 6 month intervals during the first years of life. In: Harrison DE (ed) Genetic Effects on Aging II. Telford Press, Caldwell, NJ, pp. 279–358

Davidson EH (1991) Spatial mechanisms of gene regulation in metazoan embryos. Development 113:1–26

Edelman GM (1987) Neural Darwinism. Basic Books, New York

Eichenlaub-Ritter U, Chandley AC, Gosden RG (1988). The CBA mouse as a model for age-related aneuploid in man: Studies of oocyte maturation, spindle formation, and chromosome alignment during meiosis. Chromosoma 96:220–226

Faddy MJ, Gosden RG, Edwards RG (1983) Ovarian follicle dynamics in mice: A comparative study of three inbred strains and an F1 hybrid. J Endocrinol 96:23–33

Faddy MJ, Gosden RG (1995) A mathematical model of follicle dynamics in the human ovary. Human Reproduction vol. 10 no. 4 pp 770–775

Finch CE (1982) Rodent models for aging processes in the human brain. In: Corkin S, Davis KL, Growden JH, Usdin E, Wurtman RJ (eds) Alzheimer's disease: a report of progress. Aging, Vol. 19. Raven Press, New York, pp 249–257

Finch CE (1990) Longevity, senescence, and the genome. University of Chicago Press, Chicago

Finch CE (1994) The evolution of ovarian oocyte decline with aging and possible relationships to Down syndrome and Alzheimer disease. Exp Gerontol 29:299–304

Finch CE (1996) Biological bases for plasticity during aging of individual life histories: In: Magnusson D (ed) The life span development of individuals: biological and phychosocial perspectives, a synthesis. Cambridge University Press, Cambridge, pp. 488–512

Finch CE, Pike MC (1996) Maximum lifespan predictions from the Gompertz mortality model. J Gerontol 51A:B183–194

Gavrilov LA, Gavrilova NS (1991) The biology of life span: a quantitative approach. Harwood Academic, Chur, Switzerland

Gosden RG (1985) The biology of menopause: the causes and consequences of ovarian aging. New York, Academic Press

Gosden RG, Laing SC, Felicio LS, Nelson JF, Finch CE (1983) Imminent oocyte exhaustion and reduced follicular recruitment mark the transition to acyclicity in aging C57BL/6J mice. Biol Reprod 28:255–260

Henderson G, Tomlinson BE, Gibson PH (1980) Cell counts in human cerebral cortex in normal adults throughout the life-span using an image analyzing computer. J Neurol Sci 46:113–136

Holinka CF, Tseng Y-C, Finch CE (1979) Reproductive aging in C57BL/6J mice: Plasma progesterone, viable embryos and resorption frequency throughout pregnancy. Biol Repro 20:1201–1211

Hook EB (1981) Rates of chromosomal abnormalities at different maternal ages. Obstet Gynecol 52:282–285

Jacobsen M (1985) Clonal analysis of the vertebrate CNS. TiNS 4:2856–2565

Jones EC, Krohn PL (1961) The relationships between age, numbers of oocytes, and fertility in virgin and multiparous mice. J Endocrinol 21:469–496

Jones HB (1956) A special consideration of the ageing process, disease, and life expectancy. Adv Biol Med Physics 4:281–337

Jouandet ML, Tramp MJ, Herron DM, Loftus WC, Bazell J, Gazanniga MS (1989) Brain prints: computer generated cerebral cortex in vivo. J Cogn Neurosci 1:88–117

Klein J (1975) Biology of the mouse histocompatibility-2 complex. Springer-Verlag, New York

Matsui Y, Zsebo K, Hogan BML (1992) Derivation of pluripotential embryonic stem cells from murine primordial germ cells in culture. Cell 70:841–847

Nelson JF, Felicio LS (1986) Radical ovarian resection advances the onset of persistent vaginal cornification but only transiently disrupts hypothalamic-pituitary regulation of cyclicity in C57BL/6J mice. Biol Reprod 35:957–964

Paganini-Hill A, Henderson VW (1994) Estrogen deficiency and risk of Alzheimer disease. Am J Epidemiol 140:256–261

Pesce M, Farrace MZ, Piacentini M, Dolci S, De Felici M (1993) Stem cell factor and leukemia inhibitory factor promote primordial germ cell survival by suppressing programmed cell death (apoptosis). Development 118:1089–1094

Phelan JP, Austad SN (1994) Selecting animal models of human aging: inbred stains often exhibit less biological uniformity than F1 hybrids. J Gerontol 49:B1–B11

Plant TM (1994) Puberty in primates. In: Knobil E (ed) Physiology of reproduction. Vol. 2. Raven Press, New York, pp 453–485

Rasmussen T, Schliemann T, Sorensen JC, Zimmer J, West MJ (1996) Memory impaired aged rats: No loss of principal hippocampal and subicular neurons. Neurobiol Aging 17:143–147

Richardson SJ, Senikas V, Nelson JF (1987) Follicular depletion during the menopausal transition: evidence for accelerated loss and ultimate exhaustion. J Clin Endocrinol Metab 65:1231–1240

Schneider LS, Farlow MR, Henderson VW, Pogoda JM (1996) Effects of estrogen replacement therapy on response to tacrine in patients with Alzheimer's disease. Neurology 46:1580–1584

Talman WT, Snyder D, Reis DJ (1980) Chronic lability of arterial pressure produced by destruction of A2 catecholamine neurons in rat brain stem. Circ Res 46:842–853

Upadhyay S, Zamboni L (1982) Ectopic germ cells: Natural model for the study of germ cell differentiation. Proc Natl Acad Sci 79:6584–6588

vom Saal FS, Moyer CL (1985) Prenatal effects on reproductive capacity during aging in female mice. Biol Reprod 32:1111–1126

vom Saal FS, Nelson JF, Finch CE (1994) The natural history of reproductive humans, laboratory rodents, and selected other vertebrates. In: Knobil E (ed) Physiology of reproduction. Raven Press, New York, pp 1213–1314

West MJ, Slomianka L, Gundersen HJG (1991) Unbiased stereological estimation of the total number of neurons in the subdivisions of the tat hippocampus using the optical Fractionator. Anat Rec 231:482–497

West MJ (1993) Regionally specific loss of neurons in the aging human hippocampus. Neurobiol Aging 14:287–293

Wickelgren I (1996) Is hippocampal cell death a myth? Science 271:1229–1230

Winklbauer R, Hausen P (1983) Development of the lateral line system in Xenopus laevis. 2. Cell multiplication and organ formation in the supraorbital system. J Embryol Exp Morphol 76:283–296

Subject index

Printing: Saladruck, Berlin
Binding: Buchbinderei Lüderitz & Bauer, Berlin